T0299738

Heating and Cooling of Air Through Coils

Heating and Cooling of Air Through Coils combines theory and practice to cover the fundamentals in the processes of heating and cooling of air through coils and the key aspects in the psychrometric chart, the coil fluid piping systems, the coils, and the energy sources for the fluid in the coils.

This book covers the integral elements that have a significant impact on the heating and cooling of air through coils, including the coil types, coil tube constructions and arrangements, and fluid flow characteristics in the coils. It also discusses sustainable and renewable energy sources used to heat and cool the fluid flowing in the piping system and the coils. In addition, the book covers the application of coils in central air-conditioning systems and split air-conditioning systems.

- Presents the fundamentals of heating and cooling of air through coils.
- Explains the psychrometric chart used for assessing the physical and thermodynamic properties of air in the heating and cooling processes.
- Covers numerous coil types and constructions.
- Discusses the key equipment used in the coil fluid piping systems that deliver hot water, steam, condensate, and chilled water to and from the coils.
- Considers various energy sources to the fluid in the coil piping system for heating and cooling, including solar heat energy, ocean thermal energy, and geothermal energy.

This book will interest engineers and researchers involved in the design and operation of heat exchangers and HVAC systems. It can also be used as a textbook for undergraduate and graduate students majoring in relevant fields, such as thermal fluids, HVAC, and energy management.

Heating and Cooling of Air Through Coils
Principles and Applications

Yongjian Gu

CRC Press
Taylor & Francis Group
Boca Raton London New York

CRC Press is an imprint of the
Taylor & Francis Group, an **informa** business

Designed cover image: Yongjian Gu and Marlo Heat Transfer Solutions, a Leonardo DRS brand

First edition published 2024
by CRC Press
6000 Broken Sound Parkway NW, Suite 300, Boca Raton, FL 33487-2742

and by CRC Press
4 Park Square, Milton Park, Abingdon, Oxon, OX14 4RN

CRC Press is an imprint of Taylor & Francis Group, LLC

© 2024 Yongjian Gu

Library of Congress Cataloging-in-Publication Data

Names: Gu, Yongjian, author.
Title: Heating and cooling of air through coils : principles and applications / Yongjian Gu.
Description: First edition. | Boca Raton : CRC Press, 2024. | Includes bibliographical references and index.
Identifiers: LCCN 2023002727 (print) | LCCN 2023002728 (ebook) |
ISBN 9781032266633 (hbk) | ISBN 9781032266640 (pbk) | ISBN 9781003289326 (ebk)
Subjects: LCSH: Hot-air heating. | Air conditioning. | Heat--Convection. |
Heating-pipes. | Tubes.
Classification: LCC TH7601 .G83 2024 (print) | LCC TH7601 (ebook) |
DDC 697.07--dc23/eng/20230214
LC record available at https://lccn.loc.gov/2023002727
LC ebook record available at https://lccn.loc.gov/2023002728

ISBN: 978-1-032-26663-3 (hbk)
ISBN: 978-1-032-26664-0 (pbk)
ISBN: 978-1-003-28932-6 (ebk)

DOI: 10.1201/9781003289326

Typeset in Times
by KnowledgeWorks Global Ltd.

Contents

Preface

This is a book to describe, analyze, and calculate the heating and cooling of air passing through coils. The heating and cooling of air passing through coils are the processes of heat transfer between the air and fluid flowing inside the coils. The heat transfer between the air and fluid should be balanced with the energy provided from energy sources. There are a variety of energy sources available from conventional to renewable. A coil fluid system is used to transport the fluid from the energy sources to coils. In engineering practice, various piping systems to transport fluids can be applied. These topics are covered in seven chapters: introduction, fundamentals, psychrometric chart, coils, graphic solutions of processes, coil fluid systems, and energy sources.

Chapter 1 briefly describes the heating and cooling coils and the processes of air passing through coils. The chapter introduces the tools used in analysis and calculation of the heating and cooling processes and the coil selection, such as equations, tables, diagrams, charts, and computer software.

Chapter 2 presents the fundamentals of thermodynamics, fluid dynamics, and heat transfer involved in analysis and calculation of the heating and cooling of air passing through coils. Examples in the chapter illustrate applications of the knowledge learned in this chapter for problem solving.

Chapter 3 describes the psychrometric chart. The chart is an important tool to graphically analyze and represent the psychrometric properties of moisture air, psychrometric processes, and property relationships. Examples in the chapter illustrate the determination of psychrometric properties, the manipulation of property relationships, and the identification of processes on the chart.

Chapter 4 describes the heating and cooling coils. In this chapter, various coils classified by different ways are described. Coil applications in different air-conditioning systems are also presented. Examples of selection of the heating coil and the cooling coil through calculation and by computer software are illustrated, respectively.

Chapter 5 describes the common processes of air through coils in engineering practice on the psychrometric chart. The factors involved in cooling with dehumidifying, such as bypass, sensible heat, effective sensible heat, room sensible heat, and grand sensible heat, are illustrated. Factor relationships and graphic solutions of the processes on the psychrometric chart are described. The method of air change rate used to determine the coil load is also presented in the chapter.

Chapter 6 describes the coil fluid systems and equipment and components applied in the systems. Air passing through the coil is an external flow system and fluid flowing inside the coil tubes is an internal flow system. A piping system is an internal flow system for transporting fluid to coils. Diverse coil fluid systems and various equipment and components involved in the systems are illustrated in the chapter.

Chapter 7 describes the energy sources for heating and cooling of air passing through coils. There are a variety of energy sources available for heating and cooling from conventional, such as fuel combustion and waste heat recovery to renewable, such as solar, geothermal, ocean thermal, and wind. In this chapter, various energy sources are presented and their applications in heating and cooling of air passing through coils are illustrated.

This book can be used as a textbook for undergraduates and graduate students majoring in relevant fields, such as thermal fluids, HVAC, and energy management. It also can be used as a reference for engineers and researchers who are interested applying the knowledge in this book to the design and operation of the heating and cooling of air systems.

Nomenclature

Roman

A	Flow sectional area, Surface area, Area	m^2
A_c	Cross-sectional area	m^2
A_s	Surface area	m^2
c_v	Specific heat at constant volume	J/kg.K, kJ/kg.K
c_p	Specific heat at constant pressure	J/kg.K, kJ/kg.K
D	Diameter	m
D_h	Hydraulic diameter	m
E	Energy	J, kJ
	Radiation energy	W/m^2
E_b	Black body radiation energy	W/m^2
e	Specific energy	J/kg, kJ/kg
$\dot{E}$	Energy rate, power	W, kW
f	Flow friction factor	–
g	Gravitational acceleration	m/s^2
G	Irradiation, incident radiation	W/m^2
$\dot{G}$	Airflow load	m^3/hr
	Fuel consumption rate	kg/s, kg/min
h	Convection heat transfer coefficient	W/m^2.K, kW/m^2.K
	Specific enthalpy	kJ/kg
h_a	Specific enthalpy of dry air	kJ/kg dry air
h_l	Specific latent heat	kJ/kg
h_s	Specific sensible heat	kJ/kg
h_{fg}	Specific latent enthalpy Evaporating heat	kJ/kg
h_F	Major head loss, friction head loss	m
h_L	Local head loss, minor head loss	m
h_v	Specific enthalpy of water vapor	kJ/kg
H	Enthalpy	kJ
	Flow head, pump head	m
H_{Loss}	Total head loss	m
k	Thermal conductivity	W/m.°C, W/m.K
K_L	Minor loss coefficient	–
ke	Specific kinetic energy	J/kg, kJ/kg, MJ/kg
KE	Kinetic energy	J, kJ, MJ
L_{equiv}	Equivalent length	m
L	Length, length of tube bank	m
m	Mass	kg
m_a	Mass of dry air	kg
m_v	Mass of water vapor	kg
$\dot{m}$	Mass flow rate	kg/s
Nu	Nusselt number	–

Nu_D	Nusselt number characterized by the tube outside diameter D	–
pe	Specific potential energy	J/kg, kJ/kg, MJ/kg
PE	Potential energy	J, kJ, MJ
p	Pressure	Pa, kPa, MPa
P	Circumference of surface area	m
	Pressure	Pa, kPa, MPa
	Wetted perimeter	m
P_a	Absolute pressure, pressure of dry air	Pa, kPa, MPa
P_g	Gauge pressure, gas pressure	Pa, kPa, MPa
P_v	Pressure of water vapor	Pa, kPa, MPa
q	Specific heat	J/kg, kJ/kg, MJ/kg
Q	Heat	J, kJ, MJ
$\dot{Q}$	Heat transfer rate	W, kW, J/s, kJ/s
$\dot{Q}_b$	Black body radiation energy rate	W, kW, J/s, kJ/s
$\dot{Q}_{in}$	Heat transfer rate in	W, kW, J/s, kJ/s
$\dot{Q}_{out}$	Heat transfer rate out	W, kW, J/s, kJ/s
$\dot{Q}_l$	Latent heat	W, kW, J/s, kJ/s
$\dot{Q}_s$	Sensible heat	W, kW, J/s, kJ/s
R	Gas constant	J/kg.K, J/kg.°C
	Thermal resistance	°C/W
Re	Reynolds number	–
R_h	Hydraulic radius	m
s	Specific entropy	kJ/kg.K
S_D	Diagonal pitch of tube bank	m
S_L	Longitudinal pitch of tube bank	m
S_T	Transverse pitch of tube bank	m
t	Time	s
T	Temperature	K, °C
T_{adp}	Apparatus dew-point temperature	K, °C
T_{db}	Dry-bulb temperature	K, °C
T_{dp}	Dew-point temperature	K, °C
T_{LM}	Log mean temperature difference	K, °C
T_{pp}	Pinch point temperature	K, °C
T_s	Surface temperature	K, °C
T_{wb}	Wet bulb temperature	K, °C
T_∞	Surrounding temperature	K, °C
v	Specific volume	m³/kg
V	Velocity	m/s
	Volume	m³
$\dot{V}$	Volume flow rate	m³/s
u	Specific internal energy	J/kg, kJ/kg, MJ/kg
U	Internal energy	J, kJ, MJ
	Overall heat transfer coefficient	W/m².K, kW/m².K
w	Specific work	J/kg, kJ/kg, MJ/kg
W	Work	J, kJ, MJ

$\dot{w}$	Specific power	W/kg, kW/kg, J/kg.s, kJ/kg.s
$\dot{W}$	Power	W, kW, J/s, kJ/s
z	Elevation, Height	m

Greek

ε	Emissivity of radiation	–
	Surface roughness	m
η	Efficiency	%
η_{th}	Thermal efficiency	%
ρ	Density	kg/m^3
μ	Viscosity	kg/m.s
v	Kinematic viscosity	m^2/s
σ	Stefan-Boltzmann constant	5.670×10^{-8} W/m^2.K^4
ϕ	Relative humidity	%
ω	Absolute humidity	kg H$_2$O/kg dry air
Δ	Difference	–

List of Abbreviations

AC	air conditioning
A/C	air conditioning
ACH	air change rate
ADP	apparatus dew point
AHU	air handing unit
AMTD	arithmetic mean temperature difference
BEP	best efficiency point
BF	bypass factor
CCHP	combined cooling, heating, and power
CF	contact factor
CHW	chilled water
CHWR	chilled water return
CHWS	chilled water supply
CSP	concentrating solar power
C.V.	control volume
DBT	dry-bulb temperature
ESHF	effective sensible heat factor
FCU	fan coil unit
GSHF	grand sensible heat factor
H$_2$O-LiBr	water-lithium bromide
HAWT	horizontal-axis wind turbine
HHV	high heating value of fuel
HRSG	heat recovery steam generator
HTF	heat transfer fluid
HVAC	heating, ventilation, and air conditioning
LHV	low heating value of fuel
LMTD	log mean temperature difference
NPSH	net positive suction head
OTEC	ocean thermal energy conversion
PV	photovoltaic
RLH	room latent heat
RH	relative humidity
RSH	room sensible heat
RSHF	room sensible heat factor
RTU	rooftop air handing unit
SG	specific gravity
SHF	sensible heat factor
SHR	sensible heat ratio
TCV	temperature control valve
UHTR	unit heat transfer rate
VAWT	vertical-axis wind turbine
WBT	wet-bulb temperature

About the Author

Yongjian Gu received the MSc degree and the PhD degree in mechanical engineering from the State University of New York (SUNY) at Stony Brook. He is an associate professor at the U.S. Merchant Marine Academy (USMMA) and an adjunct associate professor at New York Institute of Technology (NYIT). He teaches thermal fluids, total energy systems and design, propulsions, marine steam plants and components, gas turbines and marine auxiliary equipment, engineering economics, and other courses to undergraduates and graduate students. He holds a license of registered Professional Engineer (P.E.) in the State of New York. He also holds a certificate of professional Oracle Database Administrator (DBA). Prior to teaching at the academic institutions, he worked at industrial corporations, consulting firms, and the U.S. Department of Energy's (DOE) national laboratory as a senior mechanical engineer, project engineer, and lead engineer for many years. He has rich and valuable academic and professional experiences in HVAC systems, piping systems, thermal energy systems, and renewable/sustainable energy applications. He is actively involved in engineering research and development and academic activities. He has many publications in his subject areas of teaching and professional expertise. He serves as a peer-reviewer for multiple scientific and technical journals. He also sits on several professional boards and conference committees.

1 Introduction

1.1 BACKGROUND

A coil is a device applied to transfer thermal energy between fluids. The coil for heating purposes is called a heating coil and for cooling purposes is called a cooling coil, accordingly. Heating and cooling of air passing through coils is the process of thermal energy transfer from one fluid flowing inside the coil and the air passing outside the coil. During the heating process, the hot fluid flowing inside the coil releases heat to the air. While during the cooling process, the cold fluid flowing inside the coil takes heat from the air. After leaving the coil, the conditioned air, warm or cold, is supplied to downstream equipment or air-conditioned spaces. The technology of the heating and cooling of air passing through coils is widely applied in industry, transportation, commercial buildings, and houses. For example, in heating, ventilation, and air conditioning (HVAC) applications, the heating and cooling of air passing through the coils are common processes to provide conditioned air in spaces for comfortable living and working environments. The energy transfer mechanisms and processes of the heating and cooling of air passing through coils are complicated and depend on not only the interaction between the fluid flowing inside the coil and the air passing through the coil, but also the fluid types and flow styles in the piping systems connecting the energy sources and the coils. The description, analysis, and calculation of the heating and cooling of air passing through the coil significantly involve the knowledge of thermodynamics, fluid dynamics, and heat transfer. The selection of heating and cooling coils is also a complex task. In engineering practice, tools are widely adopted to aid the analysis and calculation for the heating and cooling of air passing through the coils and the coil selection.

1.2 TOOLS OF ANALYSIS AND CALCULATION

Many tools are applied to aid the analysis and calculation for the heating and cooling of air passing through the coils and the selection of the heating and cooling coils. Equations, tables, diagrams, charts, and software are the commonly used tools. Skillful application of these tools enables the analysis and calculation of heating and cooling of air passing through coils and the selection of the heating and cooling coils to be accomplished accurately and efficiently.

1.2.1 EQUATIONS

Equations are mathematical expressions that involve unknown variables that are uncovered by solving the equations. In engineering applications, equations depict the laws of physics and engineering phenomena. In general, there are two types of equations: explicit and implicit. Most of the equations dealt with in engineering

DOI: 10.1201/9781003289326-1

applications are the explicit ones, which can be manipulated so the unknown variable is shown on one side of the equation, such as the ideal gas equation of state in thermodynamics,

$$Pv = RT \tag{1.1}$$

and the Bernoulli equation in fluid mechanics,

$$\frac{p}{\rho} + \frac{V^2}{2} + gz = \text{constant} \tag{1.2}$$

Solving the explicit equation to determine the value of the unknown variable is not difficult. The steps involved for replacing the variables by known values and finding the value of the unknown variable are straightforward. Sometimes, the application may require implicit equations. Implicit equations have unknown variables on both the left and right sides of the equation. For instance, the Colebrook equation in fluid mechanics used to find the friction factor f in internal turbulent flows,

$$\frac{1}{\sqrt{f}} = -2.0 \log \left(\frac{\frac{\varepsilon}{D}}{3.7} + \frac{2.51}{Re\sqrt{f}} \right) \tag{1.3}$$

The equation has the unknown variable f on both sides. To solve the implicit equation, trial-and-error, graphical, or iteration methods may be applied. However, the methods may take time and effort to get the result. To get the result quickly and conveniently, values of the unknown variable in the implicit equation can be calculated in advance in all possible conditions and tabulated in tables.

1.2.2   TABLES

Tables show the values of variables visually in order according to the application conditions. The values of the variables in the tables are nicely organized and able to be found quickly. Tables are widely used in engineering practice. Some tables are created because the relationships among the variables are too complex to be expressed by equations, for example, the table of properties of saturated steam as shown in Table 1.1 and the table of specific heat of air (1 atm pressure) as shown in Table 1.2. While some tables created are simply for finding the values of the variables easily in the tabulated forms, such as the table of pipe dimension and capacity as shown in Table 1.3.

1.2.3   DIAGRAMS

Diagrams are the graphic representation to display information, data, or properties. The application of diagrams is prevalent in engineering practice for analysis that may not need to exactly show the quantitative data (numerical data), but

TABLE 1.1
Properties of Saturated Steam

P, MPa	T, °C	Density (kg/m³)		Enthalpy (kJ/kg)			Entropy (kJ/kg.K)			Volume (cm³/g)	
		ρ_l	ρ_v	h_l	h_v	Δh	s_l	s_v	Δs	v_l	v_v
0.10	99.606	958.63	0.590 34	417.5)	2,674.9	2,257.4	1.3028	7.3588	6.0561	1.043 15	1,693.9
0.15	111.349	949.92	0.862 60	467.13	2,693.1	2,226.0	1.4337	7.2230	5.7893	1.052 73	1,159.3
0.20	120.210	942.94	1.1291	504.70	2,706.2	2,201.5	1.5302	7.1269	5.5967	1.060 52	885.68
0.25	127.411	937.02	1.3915	535.34	2,716.5	2,181.1	1.6072	7.0524	5.4452	1.067 22	718.66
0.30	133.522	931.82	1.6508	561.43	2,724.9	2,163.5	1.6717	6.9916	5.3199	1.073 17	605.76
0.35	138.857	927.15	1.9077	584.26	2,732.0	2,147.7	1.7274	6.9401	5.2128	1.078 57	524.18
0.40	143.608	922.89	2.1627	604.65	2,738.1	2,133.4	1.7765	6.8955	5.1190	1.083 55	426.38
0.42	145.375	921.28	2.2642	612.25	2,740.3	2,128.0	1.7946	6.8791	5.0846	1.085 44	441.65
0.44	147.076	919.72	2.3655	619.58	2,742.4	2,122.8	1.8120	6.8636	5.0516	1.087 29	422.74
0.46	148.716	918.20	2.4666	626.64	2,744.4	2,117.7	1.8287	6.8487	5.0199	1.089 08	405.42
0.48	150.300	916.73	2.5674	633.47	2,746.3	2,112.8	1.8448	6.8344	4.9895	1.090 84	389.50
0.50	151.831	915.29	2.6680	640.39	2,748.1	2,108.0	1.8604	6.8207	4.9603	1.092 55	374.81
0.52	153.314	913.89	2.7685	646.50	2,749.9	2,103.4	1.8754	6.8075	4.9321	1.094 23	361.20
0.54	154.753	912.52	2.8688	652.72	2,751.5	2,098.8	1.8899	6.7948	4.9049	1.095 87	348.58
0.56	156.149	911.18	2.9689	658.77	2,753.1	2,094.4	1.9040	6.7825	4.8786	1.097 48	336.82
0.58	157.506	909.87	3.0689	664.65	2,754.7	2,090.0	1.9176	6.7707	4.8531	1.099 05	325.85
0.60	158.826	908.59	3.1687	670.38	2,756.1	2,085.8	1.9308	6.7592	4.8284	1.100 60	315.58
0.62	160.112	907.34	3.2684	675.96	2,757.6	2,081.6	1.9437	6.7482	4.8045	1.102 12	305.96
0.64	161.365	906.11	3.3680	681.41	2,758.9	2,077.5	1.9562	6.7374	4.7813	1.103 62	296.91
0.66	162.587	904.91	3.4675	686.73	2,760.3	2,073.5	1.9684	6.7270	4.7587	1.105 09	288.40
0.68	163.781	903.72	3.5668	691.92	2,761.5	2,069.6	1.9802	6.7169	4.7367	1.106 54	280.36

TABLE 1.2

Specific Heat of Air (1 atm Pressure)

Temperature T (K)	Gas Constant R (kJ/kg.K)	Density ρ (kg/m³)	Specific Heat at Constant Pressure c_p (kJ/kg.K)	Specific Heat at Constant Volume c_v (kJ/kg.K)	Specific Heat Ratio k
240	0.2870	1.4710	1.006	0.7164	1.404
260		1.3579	1.006	0.7168	1.403
273.2		1.2923	1.006	0.7171	1.403
280		1.2609	1.006	0.7173	1.402
288.7		1.2229	1.006	0.7175	1.402
300		1.1768	1.006	0.7180	1.402
320		1.1033	1.007	0.7192	1.400
340		1.0384	1.009	0.7206	1.400
360		0.9807	1.010	0.7223	1.398
380		0.9291	1.012	0,7243	1.397
400		0.8826	1.014	0.7266	1.396

TABLE 1.3

Pipe Dimension and Capacity

Diameter Nominal D_n (mm)	Outside Diameter D (mm)	Wall Thickness t (mm)	Inside Diameter d (mm)	Sectional Area A_i (cm²)	Inside Volume V (cm³/m)	Inside Water Mass m (kg/m)	Inside Water Weight W (N/m)
15	21.34	2.769	15.798	1.960	196.00	0.196	1.923
20	26.67	2.870	20.930	3.441	344.10	0.344	3.376
25	33.40	3.378	26.645	5.576	557.60	0.558	5.470
32	42.16	3.556	35.052	9.650	965.00	0.965	9.467
40	48.26	3.683	40.894	13.134	1,313.40	1.313	12.884
50	60.33	3.912	52.501	21.648	2,164.80	2.165	21.237
65	73.03	5.156	62.713	30.889	3,088.90	3.089	30.302
80	88.90	5.486	77.928	47.695	4,769.50	4.770	46.789
90	101.60	5.740	90.120	63.787	6,378.70	6.379	62.575
100	114.40	6.020	102.260	82.130	8,213.00	8.213	80.570
125	141.30	6.553	128.194	129.070	12,907.00	12.907	126.618
150	168.28	7.112	154.051	186.388	18,638.80	18.639	182.847

* Water density $\rho = 1,000$ kg/m³

rather to represent the variable relations, variable development tendency, conceptual explanation, or abstract information. There are three major types of diagrams commonly used in thermal fluids application: the mechanical system diagram, flow diagram, and property diagram. The mechanical system diagram is a schematic drawing used to describe the interconnection of components in the

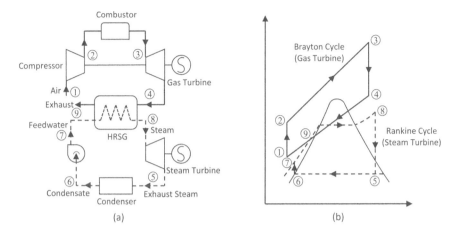

FIGURE 1.1   Diagrams of a gas turbine-based combined cycle power plant. (a) System diagram. (b) Property diagram.

mechanical system. The flow diagram is a schematic drawing with appropriate standard symbols and connecting lines used to define the fluid flow sequence. In the flow diagram, mechanical equipment, components, connections, gauges, etc., are shown. The property diagram is a schematic drawing used to represent the thermodynamic properties and states of thermal fluids and the consequences of the fluid through certain processes. The exact values of the properties in the property diagrams, in general, may not need to be shown. Figure 1.1(a) and (b) show the mechanical system diagram and the property diagram of a gas turbine-based combined cycle power plant, respectively. Figure 1.2 shows the airflow diagram of an HVAC system. Figure 1.3 is a diagram showing the temperature pinch point of the fluids in a heat recovery steam generator (HRSG). The diagram is quite useful to analyze the relationships between the energies and the temperatures of the fluids even though their quantitative values are not represented.

1.2.4 CHARTS

Charts are graphical formats used to display a series of numeric data. The chart makes the data and its relationships more easily understood. The data in the chart may be represented by symbols, such as bars in a bar chart, lines in a line chart, or slices in a pie chart. The chart may represent tabular numeric data, functions, or some kind of quality structures. Some charts are created, just like tables, only because the relationships among the variables are too complex to be expressed by the equations. Figure 1.4 shows the chart of U.S. renewable energy consumption in 2019. Figures 1.5 and 1.6 show the Moody chart (friction factor f of internal turbulent flow) and the psychrometric chart at a pressure of 1 atm (101.325 kPa), respectively. The moody chart and the psychrometric chart are quite useful in the analysis and calculation of the heating and cooling of air passing through coils.

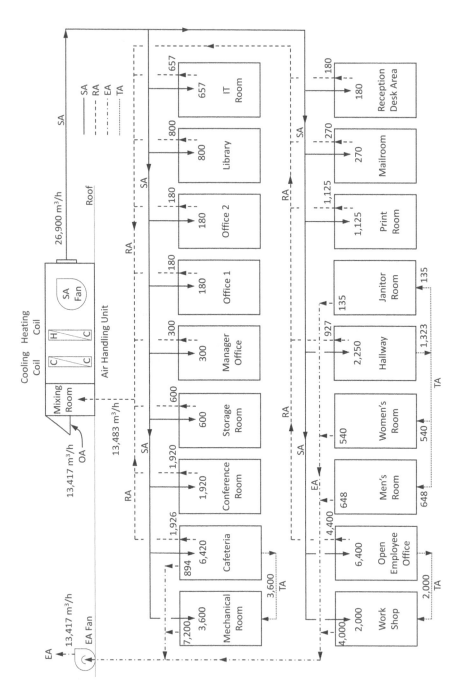

FIGURE 1.2   The air flow diagram of an HVAC system.

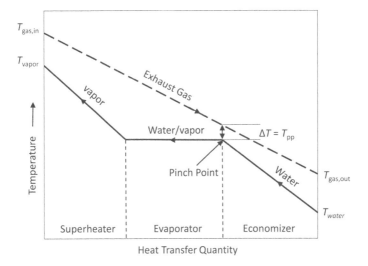

FIGURE 1.3 The temperature pinch point in an HRSG.

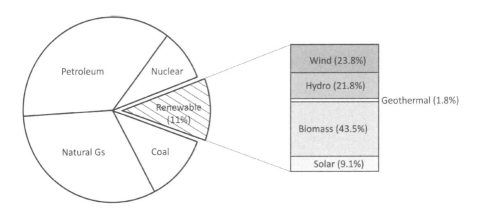

FIGURE 1.4 U.S. renewable energy consumption in 2019. (Adapted from Center for Sustainable Systems, University of Michigan, U.S. September 2021.)

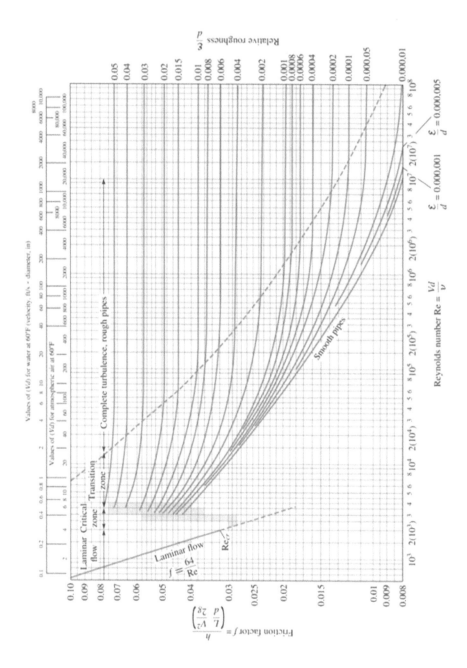

FIGURE 1.5   The Moody chart (friction factor f of internal turbulent flow).

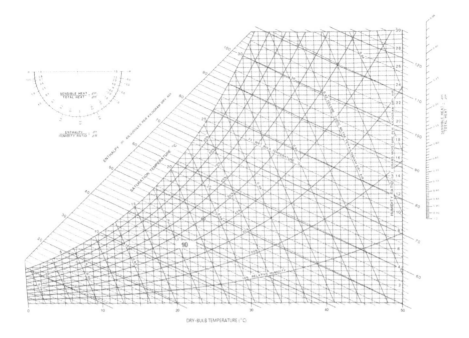

FIGURE 1.6 Psychrometric chart at a pressure of 1 atm (101.325 kPa).

1.2.5 SOFTWARE

Software is a computer program that performs a specific function. In engineering practices, software plays an important role. Software applications help save time and avoid the difficulty of conducting large and complex projects. There are different kinds of software used widely in engineering practices, such as for calculation, analysis, or equipment selection. Software applications are computer based. Some software can be downloaded from the Internet, for instance, the engineering equation solver (EES) as shown in Figure 1.7. Some software may be available only for online use, for instance, the online unit convertor as shown in Figure 1.8. Some software is used for a company's product selection, for instance, the coil selection software as shown in Figure 1.9.

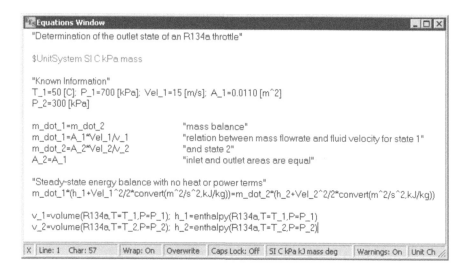

FIGURE 1.7 The engineering equation solver (EES). (Extracted from F-Chart Software, LLC. Engineering Equation Solver, 2022.)

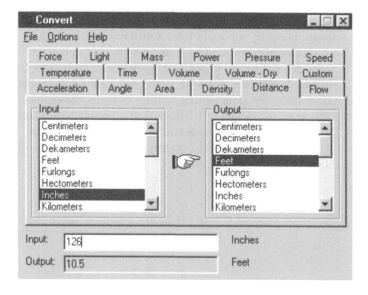

FIGURE 1.8 The unit converter. (Extracted from joshmadison.com, 2022.)

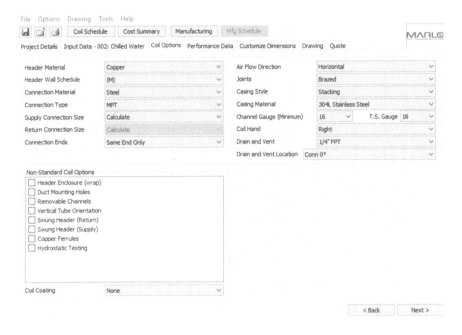

FIGURE 1.9   The coil selection software. (Extracted from Marlo Heat Transfer Solutions, coil and AHU selection software, 2022.)

2 Fundamentals of Heating and Cooling

2.1 BASICS OF THERMODYNAMICS

Thermodynamics is the study of the relationships between heat, work, and energy exchange. Thermodynamics is an important subject to define pressure, temperature, and states of thermodynamic systems. The basic knowledge of the thermodynamic system and the system properties is fundamental to the study of heating and cooling processes, such as the heating and cooling of air passing through coils.

2.1.1 THERMODYNAMIC SYSTEMS

A thermodynamic system is defined as a quantity of matter in a region or space chosen for study. The area outside the system is called surroundings. A real or imaginary surface separating the system and its surroundings is called boundary. A schematic of a thermodynamic system, its boundary, and surroundings is shown in Figure 2.1.

A thermodynamic system can be in a closed, open, or isolated style. The style simply depends on whether the mass and energy in the system can be exchanged with the surroundings. A closed system has fixed mass within the boundary and cannot have mass exchange with the surroundings. Energy in the form of heat or work, however, can cross the boundary of the closed system and have exchange with the surroundings. An open system can have mass and energy crossing the boundary and make exchange of both with the surroundings. Since a space or region of an open system for study involves a mass flow in and out, a control volume (C.V.) is usually applied to specify the space or region. If no mass and energy are allowed to cross the boundary, such a system is called an isolated system. Figure 2.2 shows a typical closed system and open system, respectively.

System Properties

Any characteristic of a thermodynamic system is called a system property. Some familiar system properties are pressure P, temperature T, volume V, and mass m. Properties can be categorized as either intensive or extensive. The intensive properties are independent of the mass of a system, such as temperature, pressure, and density. The extensive properties are dependent on the size of a system, such as total mass, total volume, and total momentum. Extensive properties per unit mass are usually called specific properties, for example, specific volume ($v = V/m$) and specific energy ($e = E/m$).

DOI: 10.1201/9781003289326-2

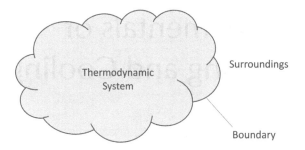

FIGURE 2.1   A thermodynamics system, its boundary, and surroundings.

State

The state that thermodynamics deals with is the system equilibrium state. The word "equilibrium" implies a stable condition with uniform properties within the system. For instance, a thermodynamic system under a thermal equilibrium means the temperature has no change in the system and is the same throughout the entire system at the time. The state of a system is defined by its properties. But it is not necessary to know all properties of the system to identify the system state. Once a sufficient number of properties are known, the state of a system can be identified and the rest of the properties can be confirmed by the state of the system. The number of properties required to identify the state of a system containing a pure substance is defined by the state postulate described as:

> *The state of a system containing a pure substance may be uniquely specified by knowing its two independent, intensive properties.*

In an engineering application, two independent properties, such as temperature and pressure, often need to be specified. The system state then is identified by these two

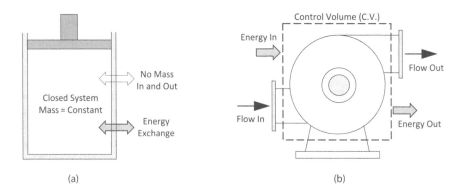

FIGURE 2.2   A typical closed system and open system. (a) A closed system (piston-cylinder device). (b) An open system (pump).

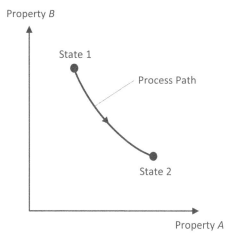

FIGURE 2.3   The relationship between the states and the process.

properties. Other properties of the system can be determined from the state by using tables, equations, or charts.

Process

Any change that a system undergoes from one equilibrium state to another is called a process. A series of states that the system passes through during a process is called the path of the process. To completely describe a process, the initial and final states, the path of the process follows, and the interactions of the system with the surroundings should be specified. Figure 2.3 shows the relationship between the states and the process on a property diagram.

2.1.2   PURE SUBSTANCES

A substance that has a fixed chemical composition throughout is called a **pure substance.** Water, for example, is a pure substance. A mixture of various chemical elements or compounds also qualifies as a pure substance as long as the mixture is homogeneous. Air, for example, is a pure substance because gases in the air generally have uniform chemical composition. A mixture of two or more phases of a pure substance is considered as a pure substance too as long as the chemical composition of all phases is the same, for example, a saturated mixture of water and steam in two phases. The two-dimensional P-v diagram and T-s diagram of a pure substance are mostly used to predict the system state and phase change and conduct the process analysis. Figure 2.4 shows the P-v diagram and T-s diagram of a pure substance, respectively.

Particularly, for the property identification, state specification, and process analysis of air, a two-dimensional diagram called psychrometric chart is commonly used. The psychrometric chart at 1 atm (101.325 kPa) is found in Appendix A.4.

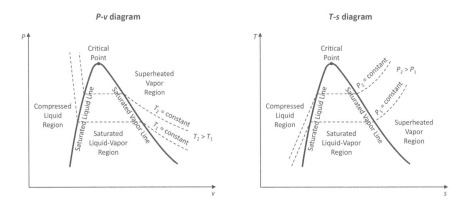

FIGURE 2.4   *P-v* and *T-s* diagrams of a pure substance.

Phase Change

A substance can have three fundamental phases: solid, liquid, and gas. Under a certain condition, a substance may appear in a different phase or undergo a phase change from one phase to another. The heat during the process in the phase change is called latent heat. During a process of phase change in a two-phase state, the substance temperature will not change when the substance continuously obtains heat or releases heat. Latent heat plays an important role in heat transfer process, which is determined by the equation

$$\dot{Q}_l = \dot{m}h_{fg} = \rho\dot{V}h_{fg} \tag{2.1}$$

When a substance is in a single phase state without phase change, adding or releasing heat will cause change in the temperature of substance. The heat during the process in the single phase is called sensible heat, which is determined by the equation

$$\dot{Q}_s = \dot{m}\Delta h_s = \dot{m}c_p\Delta T = \rho\dot{V}c_p\Delta T \tag{2.2}$$

The total heat during a process for a substance experiencing a phase change from one single-phase state to another single-phase state, therefore, is

$$\dot{Q} = \dot{Q}_s + \dot{Q}_l = \dot{m}\left(\Delta h_s + h_{fg}\right) \tag{2.3a}$$

or

$$\dot{Q} = \dot{m}\left(h_2 - h_1\right) \tag{2.3b}$$

where h_1 and h_2 are the substance enthalpies of the initial state and the final state, respectively.

Example 2.1

Water at a rate of 50 kg/s enters a boiler at a pressure of 6.0 MPa and a temperature of 60°C. The water gets heat in the boiler and becomes steam. If a superheated steam flow exits the boiler at a pressure of 6.0 MPa and a temperature of 500°C, determine the total heat rate (kW) required in the process to change the water to the superheated steam in the boiler by using *(a)* Equation (2.3a) and *(b)* Equation (2.3b).

SOLUTION

The states of the water entering the boiler and the steam exiting the boiler are specified by the given pressure and temperature. Referring to Appendix A.5 Properties of Steam and Compressed Water, the fluid at 6.0 MPa and 60°C is a compressed water in single phase. The fluid at 6.0 MPa until to 275.585°C is a water-steam mixture in two phases. The fluid at 6.0 MPa and temperature of 500°C is a superheated steam in single phase.

The heat rates of the fluid at 6.0 MPa are

$$\text{Compressed water } \Delta h_s = h_{@275.585°C} - h_{@60°C} = (1,213.9 - 256.2)\frac{kJ}{kg} = 957.7\frac{kJ}{kg}$$

$$\text{Water} - \text{steam } h_1 = h_{fg@275.585°C} = 1,570.7\frac{kJ}{kg}$$

$$\text{Superheated steam } \Delta h_s = h_{@500°C} - h_{@275.585°C} = 3,423.1 - 2,784.6 = 638.50\frac{kJ}{kg}$$

(a) Using Equation (2.3a), the total heat rate required in the process is

$$\dot{Q} = \dot{Q}_s + \dot{Q}_l = \dot{m}(\Sigma\Delta h_s + h_1)$$

$$= \left(50\frac{kg}{s}\right)(957.7 + 638.50 + 1,570.7)\frac{kJ}{kg} = 158,345 \text{ kW}$$

(b) Referring to Appendix A.5 Properties of Steam and Compressed Water, the enthalpies of the fluid entering and exiting the boiler at 6.0 MPa are determined to be

$$h_1 = h_{@60°C} = 256.2\frac{kJ}{kg}$$

$$h_2 = h_{@500°C} = 3,423.1\frac{kJ}{kg}$$

Using Equation (2.3b), the total heat rate required in the process, therefore, is

$$\dot{Q} = \dot{m}(h_2 - h_1) = \left(50\frac{kg}{s}\right)(3,423.1 - 256.2)\frac{kJ}{kg}$$

$$= 158,345 \text{ kW}$$

2.1.3   THE IDEAL GAS EQUATION OF STATE

Any equation that relates the pressure, temperature, and specific volume of a substance is called an equation of state. The simplest and most well-known equation of
state for gases is the ideal gas equation of state. The ideal gas equation of state is
defined by Equation (1.1) (see Chapter 1) as

$$Pv = RT \tag{1.1}$$

where R is a gas constant whose value depends on the gas type. Since $V = mv$, then

$$PV = mRT \tag{2.4}$$

For a fixed mass going on a process, the properties of an ideal gas at two states are
related by

$$\frac{P_1 V_1}{T_1} = \frac{P_2 V_2}{T_2} \tag{2.5}$$

Example 2.2

A vehicle tire has air inside at gauge pressure of 230 kPa and temperature 27°C
before a trip and 235 kPa after the trip, where the atmospheric pressure is 95 kPa.
If the tire remains a constant volume 0.42 m³ without any air leaking during the
trip, determine *(a)* the air temperature (°C) and *(b)* the air mass (kg) in the tire after
the trip.

SOLUTION

The state of the air in the tire is specified by the given pressure and temperature.
Referring to Appendix A.3 Specific Heat of Air, the gas constant R of air is 0.287
kJ/kg·K. And referring to Appendix A.1 Unit Conversion, 0.287 kJ/kg.K is equal to
0.287 kPa.m³/kg.K. The absolute pressures in the tire before and after the trip are
known,

$$P_1 = P_{g1} + P_a = 230 + 95 = 325 \text{ kPa}$$

$$P_2 = P_{g2} + P_a = 235 + 95 = 330 \text{ kPa}$$

(a) Using Equation (2.5), the temperature after the trip is

$$T_2 = T_1\left(\frac{P_2}{P_1}\right) = (27 + 273)\ \text{K}\left(\frac{330\ \text{kPa}}{325\ \text{kPa}}\right)$$

$$= 300\ \text{K}(1.0154) = 304.62\ \text{K} = \textbf{31.62°C}$$

(b) Using Equation (2.4), the mass in the tire after the trip is determined to be

$$m = \frac{P_2 V}{R T_2} = \frac{(330\ \text{kPa})(0.42\ \text{m}^3)}{\left(0.287\ \dfrac{\text{kPa.m}^3}{\text{kg.K}}\right)(304.62\ \text{K})} = \textbf{1.59 kg}$$

2.1.4 ENERGY, HEAT, WORK, AND EFFICIENCY

Forms of Energy

Energy can exist in different forms, such as thermal, mechanical, kinetic, potential, electric, magnetic, chemical, and nuclear. The sum of the energy in the different forms is called the total energy E. In the absence of magnetic, electric, chemical, nuclear, and surface tension, the total energy of a system only consists of the internal energy U, the kinetic energy KE, and the potential PE, i.e.,

$$E = U + KE + PE \tag{2.6}$$

and

$$KE = m\frac{V^2}{2} \tag{2.7}$$

$$PE = mgz \tag{2.8}$$

where z is the elevation of the system to a selected reference level. A specific energy, which is the total energy on a unit mass basis, is expressed as

$$e = u + ke + pe \tag{2.9a}$$

i.e.,

$$e = u + \frac{V^2}{2} + gz \tag{2.9b}$$

For an open system, an energy flow rate $\dot{E}$ is commonly used. The energy flow rate is determined by the specific energy plus a flow energy PV associated with a mass flow rate $\dot{m}$ flowing through the cross-sectional area of the control volume,

$$\dot{E} = \dot{m}(e + PV) = \dot{m}\left(u + \frac{V^2}{2} + gz + PV\right) \tag{2.10}$$

The flow energy is applied to the fluid entering or leaving a control volume. The energy combining the flow work and the internal energy is called enthalpy h,

$$h = u + Pv \tag{2.11}$$

In open flows, h is an important and useful property. Equation (2.10) then becomes

$$\dot{E} = \dot{m}\left(h + \frac{V^2}{2} + gz\right) \tag{2.12}$$

If a volume flow rate $\dot{V}$ is given, the mass flow rate can be obtained by

$$\dot{m} = \frac{\dot{V}}{v} = \rho\dot{V} = \rho A V_{avg} \tag{2.13}$$

where A is the flow cross-sectional area and V_{avg} is the average flow velocity normal to the area.

Energy Transfer by Heat

Heat Q is defined as the form of energy transferred driven by a temperature difference. Heat is not a system property, but a process function measured during a system going from one state to another. The heat transfer per unit mass of a system during a process is expressed as

$$q = \frac{Q}{m} \tag{2.14}$$

The heat can be either a sensible heat or a latent heat.

Energy Transfer by Work

Work W, like heat, is an energy form measured during a process. Work is the energy transfer associated with a force acting through a distance. Work can be simply recognized as an energy interaction that is not caused by a temperature difference. The work per unit mass of a system is called the specific work expressed as

$$W = \frac{W}{m} \tag{2.15a}$$

The work per unit time is called power expressed as

$$\dot{W} = \frac{W}{t} \tag{2.15b}$$

The specific power is

$$\dot{W} = \frac{\dot{W}}{\dot{m}} \tag{2.15c}$$

Energy Conversion Efficiency

Energy can be converted from one form to another. Efficiency η is a measurement to see how well an energy conversion or transfer process is accomplished. Efficiency is the ratio of the desired net energy output from a system to the required total energy input to the system,

$$\eta = \frac{\text{Desired net energy output}}{\text{Required total energy input}} \tag{2.16}$$

During the energy conversion, energy loss exists. Therefore, the desired net energy output is always less than the required total energy input, i.e.,

$$\eta < 1$$

In a flow system, it is usually interested to increase the desired pressure, velocity, or elevation of a fluid. This is done by supplying mechanical energy to the fluid by a pump, a fan, or a compressor. For a pump, the pump efficiency can be expressed as

$$\eta_{\text{pump}} = \frac{\text{Mechanical energy incrase of the fluid}}{\text{Mechanical energy input}}$$

$$= \frac{\Delta \dot{E}_{\text{mech,fluid}}}{\dot{W}_{\text{shaft,in}}} = \frac{\dot{W}_{\text{pump,u}}}{\dot{W}_{\text{pump}}} \tag{2.17}$$

where $\Delta \dot{E}_{\text{mech,fluid}} = \dot{E}_{\text{mech,out}} - \dot{E}_{\text{mech,in}}$ is the rate of increase in the mechanical energy of the fluid, which is equivalent to the useful pumping power $\dot{W}_{\text{pump,u}}$ supplied to the fluid. In an engineering application, a pump or a fan is usually packaged together with a motor. Therefore, there is an overall efficiency to express the efficiency of the pump-motor unit,

$$\eta_{\text{overall}} = \eta_{\text{pump-motor}} = \eta_{\text{pump}} \eta_{\text{motor}}$$

$$= \frac{\dot{W}_{\text{pump,u}}}{\dot{W}_{\text{motor}}} = \frac{\Delta \dot{E}_{\text{mech,fluid}}}{\dot{W}_{\text{motor}}} \tag{2.18}$$

where $\dot{W}_{\text{motor}}$ is the electrical consumption of the motor.

2.1.5 THE FIRST LAW OF THERMODYNAMICS

The first law of thermodynamics is a description of energy conservation. The law states that energy must remain constant during a process. In the process, the energy cannot be created or destroyed; but can change between forms. The first law of thermodynamics indicates that the net change (increase or decrease) of the total energy in a system ΔE_{system} during a process is equal to the difference of the total energy entering the system ΔE_{in} and the total energy leaving the system ΔE_{out}.

Energy balance for any system undergoing any kind of process is expressed as

$$\Delta E_{in} - \Delta E_{out} = \Delta E_{system} \tag{2.19a}$$

or

$$E_{in} - E_{out} = \Delta E_{system} \tag{2.19b}$$

In the equation, the left side is the net energy transfer between the system and surroundings and the right side is the energy change in the system itself.

In an engineering practice, most machines and components involved in applications are open systems in steady-flow, such as boilers, chillers, pumps, etc. Therefore, mass flow and energy transfer in and out across a machine or a component should be equal. The mass flow for a steady-flow system is,

$$\sum_{in} \dot{m} = \sum_{out} \dot{m} \tag{2.20}$$

The symbol of summation means there are multiple flow streams entering and leaving the control volume. The mass balance of a single-stream for a steady flow system becomes

$$\dot{m}_1 = \dot{m}_2 \tag{2.21a}$$

or

$$\rho_1 \dot{V}_1 A_1 = \rho_2 \dot{V}_1 A_2 \tag{2.21b}$$

The rate form of the energy balance for an open system is

$$\dot{E}_{in} - \dot{E}_{out} = \frac{dE_{system}}{dt} \tag{2.22}$$

In the equation, the left side is the net energy transfer rate with the surroundings, such as heat, work, and mass. The right side of the equation is the energy change rate of the system within the control volume, such as internal, kinetic, potential energy, etc. For a steady flow system, the energy change in the control volume should be zero,

$$\frac{dE_{system}}{dt} = 0$$

The energy rate entering and leaving a control volume, therefore, must be equal

$$\dot{E}_{in} = \dot{E}_{out} \tag{2.23}$$

Equation (2.23) can be written explicitly in the energy forms of heat, work, flow work, kinetic energy, and potential energy as

$$\dot{Q}_{in} + \dot{W}_{in} + \sum_{in} \dot{m}\left(h + \frac{V^2}{2} + gz\right)_{in} = \dot{Q}_{out} + \dot{W}_{out} + \sum_{out} \dot{m}\left(h + \frac{V^2}{2} + gz\right)_{out} \tag{2.24}$$

For an open system with a single-stream flow, the steady-flow energy balance equation becomes

$$\dot{Q} - \dot{W} = \dot{m}\left[h_2 - h_1 + \frac{V_2^2 - V_1^2}{2} + g(z_2 - z_1)\right]$$ (2.25)

In Equation (2.25), the left side is the rate of energy transfer between the system and surroundings in the forms of heat and work. The right side of the equation is the energy change in the system. Dividing Equation (2.25) by the mass flow rate $\dot{m}$, an energy balance per unit mass is obtained,

$$q - w = h_2 - h_1 + \frac{V_2^2 - V_1^2}{2} + g(z_2 - z_1)$$ (2.26)

When the kinetic and potential energies are negligible, i.e., $\Delta ke = \Delta pe = 0$, the energy balance per unit mass of an open system is simplified as

$$q - w = h_2 - h_1$$ (2.27)

Example 2.3

A fan in a heater with an electric resistor installed in a duct forces air to pass through the resistor. The air enters the heater at 100 kPa and 18°C and leaves the heater at 36°C with the same pressure. If the electric resistor consumes 1,600 W power and the duct cross-sectional area is 80 cm², determine (a) the mass flow rate (kg/s) of the air at the inlet, (b) volume flow rate (m³/s) of the air at the inlet, and (c) the velocity (m/s) of the air at the exit of the fan heater without considering the heat loss through the duct.

SOLUTION

Referring to Appendix A.3 Specific Heat of Air, the gas constant of air R is 0.287 kJ/kg.k and c_p is 1.006 kJ/kg.K at the given temperature. And referring to Appendix A.1 Unit Conversion, 0.287 kJ/kg.K is equal to 0.287 kPa.m³/kg.K. The control volume is the duct section including the fan heater.

(a) The air flow is a single-stream steady flow. Using Equation (2.21a) and Equation (2.23),

$$\dot{m}_1 = \dot{m}_2 = \dot{m}$$

$$\dot{E}_{in} = \dot{E}_{out}$$

When KE and PE are neglected, the energy balance becomes

$$\dot{W}_{heater} + \dot{m}h_1 = \dot{m}h_2$$

or

$$\dot{W}_{heater} = \dot{m}(h_2 - h_1) = \dot{m}c_p(T_1 - T_2)$$

The mass flow rate of the air is determined to be

$$\dot{m}_1 = \frac{\dot{W}_{heater}}{c_p(T_1 - T_2)} = \frac{1.6\,\frac{kJ}{s}}{\left(1.006\,\frac{kJ.kg}{K}\right)(36 - 18)°C} = 0.0884\,\frac{kg}{s}$$

(b) The specific volume of the air entering the fan heater is

$$v_1 = \frac{RT_1}{P_1} = \frac{\left(0.287\,\frac{kPa.m^3}{kg.K}\right)(18 + 273\ K)}{100\ kPa} = 0.8352\,\frac{m^3}{kg}$$

Using Equation (2.13), the volume flow rate of the air entering the fan heater is

$$\dot{V}_1 = \dot{m}v_1 = \left(0.0884\,\frac{kg}{s}\right)\left(0.8352\,\frac{m^3}{kg}\right) = 0.0738\,\frac{m^3}{s}$$

(c) Using the ideal gas equation of state,

$$v_2 = \frac{RT_2}{P_2} = \frac{\left(0.287\,\frac{kPa.m^3}{kg.K}\right)(36 + 273\ K)}{100\ kPa} = 0.8868\,\frac{m^3}{kg}$$

then the exit velocity of the air is determined to be

$$V_2 = \frac{\dot{m}v_2}{A_2} = \frac{\left(0.0884\,\frac{kg}{s}\right)\left(0.8868\,\frac{m^3}{kg}\right)}{80 \times 10^{-4} m^2} = 9.7991\,\frac{m}{s}$$

2.2   BASICS OF FLUID MECHANICS

Fluid mechanics is the study of the behavior of fluids at rest, in motion, and their interaction with the boundaries, such as internal flow and external flow. In heating and cooling processes of air passing through coils, internal flow is the liquid moving inside the coil and the external flow is the air passing through outside the coil. Fluid mechanics plays an important role to analyze and calculate the heating and cooling processes of air passing through coils.

2.2.1 PROPERTIES OF FLUIDS

Except the familiar properties, such as pressure P, temperature T, volume V, and mass m, of a system, some other properties are frequently used in the study of a fluid system, for example, density, specific gravity, and specific heat.

Density

Density ρ is defined as mass per unit volume and expressed as

$$\rho = \frac{m}{V} \tag{2.28a}$$

Density can also be obtained from the reciprocal of specific volume v, i.e.,

$$\rho = \frac{1}{v} \text{ or } v = \frac{1}{\rho} \tag{2.28b}$$

Specific Gravity

Sometimes the density of a substance is given by the ratio of the density of a substance to the density of a well-known standard substance at a specified condition. The ratio is called specific gravity (SG). Conventionally, water is used as the standard substance at the condition of 4°C and 1 atm. The SG of a fluid, therefore, is expressed as

$$SG = \frac{\rho}{\rho_{H_2O}} \tag{2.29}$$

where ρ_{H_2O} is 1,000 kg/m^3 at the standard condition of 4°C and 1 atm.

Specific Heats

Specific heat c is defined as the quantity of energy required to raise a unit mass of a substance by temperature in one degree. In an open system, a specific heat at constant pressure c_p is typically used, which is expressed as

$$c_p = \left(\frac{\partial h}{\partial T} \right)_p \tag{2.30}$$

The subscription p means the process is undergoing a constant pressure. The integration of Equation (2.30) yields

$$c_p dT = dh \tag{2.31a}$$

or

$$c_p \left(T_1 - T_2 \right) = h_1 - h_2 \tag{2.31b}$$

Example 2.4

Air is ventilated at a pressure of 100 kPa and a temperature of 25°C into a room whose dimensions are 10 m, 6 m, and 5 m (length, width, and height). Determine (a) the density, (b) the specific volume, (c) the specific gravity, and (d) the mass of the air.

SOLUTION

The state of the air in the room is specified by the given pressure and temperature. Referring to Appendix A.3 Specific Heat of Air, the gas constant of air R is 0.287 kJ/kg·K. And referring to Appendix A.1 Unit Conversion, 0.287 kJ/kg·K is equal to 0.287 kPa.m^3/kg.K.

(a) Using the ideal gas equation of state, the density of the air is determined as

$$\rho = \frac{P}{RT} = \frac{100 \text{ kPa}}{\left(0.287 \frac{\text{kPa.m}^3}{\text{kg.K}}\right)(25+273)\text{K}} = \mathbf{1.1692 \text{ kg}/\text{m}^3}$$

(b) The specific volume of the air is the reciprocal of the density,

$$v = \frac{1}{\rho} = \frac{1}{1.1692 \frac{\text{kg}}{\text{m}^3}} = \mathbf{0.8553 \text{ m}^3/\text{kg}}$$

(c) Using Equation (2.29), the specific gravity of the air is determined to be

$$SG = \frac{\rho}{\rho_{H_2O}} = \frac{1.1692 \frac{\text{kg}}{\text{m}^3}}{1,000 \frac{\text{kg}}{\text{m}^3}} = \mathbf{1.1692 \times 10^{-3}}$$

(d) The volume of the room is

$$V = (10 \text{ m})(6 \text{ m})(5 \text{ m}) = 300 \text{ m}^3$$

The mass of the air in the room, therefore, is

$$m = \rho V = \left(1.1692 \frac{\text{kg}}{\text{m}^3}\right)(300 \text{ m}^3) = \mathbf{350.76 \text{ kg}}$$

2.2.2   THE BERNOULLI EQUATION

The Bernoulli equation establishes relation of a fluid flow among pressure, velocity, and elevation. It is valid in regions of steady and incompressible flows where net frictional forces are negligible and without heat transfer and shaft work involved. The Bernoulli equation is one of the powerful equations in fluid mechanics. The Bernoulli equation has a basic energy form as shown in Equation (1.2) (see Chapter 1)

$$\frac{P}{\rho} + \frac{v^2}{2} + gz = \text{constant} \tag{1.2}$$

The unit of the equation in the energy form is kJ/kg. The terms of P/ρ, $v^2/2$, and gz represent the flow energy, the kinetic energy, and the potential energy, respectively. The Bernoulli equation states that the sum of the kinetic, potential, and flow energy of an incompressible fluid is a constant during a steady flow with negligible friction. This is equivalent to the energy conservation of the first law of thermodynamics. The energy terms in a fluid flow are exchangeable, but the sum should remain a constant. Dividing g to each term, the Bernoulli equation becomes a head form as

$$\frac{P}{\rho g} + \frac{V^2}{2g} + z = H = \text{constant} \tag{2.32}$$

The unit of the equation in the head form is m. The terms of $P/\rho g$, $V^2/2g$, and z represent the pressure head, the velocity head, and the elevation head, respectively. The sum of the pressure head, the velocity head, and elevation head forms a total head H.

2.2.3 THE EXTENDED BERNOULLI EQUATION

Equation (2.32) has two main restrictions:

1. No work is presented. In engineering applications, work is mostly involved, such as pump work in a hydraulic system.
2. No friction is involved. In reality, friction always occurs in a flow, which results in a head loss.

To take into account the work or head, Equation (2.32) is modified. The modified Bernoulli equation is called the extended Bernoulli equation, which is expressed as

$$\frac{P_1}{\rho g} + \frac{1}{2}\frac{v_1^2}{g} + z_1 + H_{\text{pump}} = \frac{P_2}{\rho g} + \frac{1}{2}\frac{v_2^2}{g} + z_2 + H_{\text{turbine}} + H_{\text{loss}} \tag{2.33}$$

Equation (2.33) is applied for solving most fluid flow problems in engineering practice. In the equation, the pump head H_{pump} presented on the left side is an energy source and the total head loss H_{loss} presented on the right side is an energy consumption. H_{turbine} is the turbine head extracted from the fluid flow if a turbine is involved in the flow system. The extended Bernoulli equation states that the total head loss H_{loss} of a fluid flowing in a hydraulic system should be balanced by the supplied pump head H_{pump} if the turbine is not involved. The total head loss occurring in the hydraulic system depends on many factors, such as flow pattern, fluid velocity, roughness of the pipe surface, pipe diameter, pipeline length, and fittings.

Reynolds Number

Reynolds number, Re, is used to predict the flow pattern of a fluid flow. The Re is the ratio of inertial forces to viscous forces within a fluid due to different fluid velocities,

$$\text{Re} = \frac{\text{Inertial forces}}{\text{Viscous forces}} = \frac{V_{avg}D}{v} = \frac{\rho v_{avg}D}{\mu} \tag{2.34}$$

where v and μ are called kinematic viscosity (m²/s) and dynamic viscosity (kg/m.s), respectively. They have a relation of $v = \mu/\rho$. At low Reynolds numbers, fluid flows tend to be dominated by laminar patterns. While at high Reynolds numbers, flows tend to be dominated by turbulent patterns.

Laminar Flow and Turbulent Flow

Laminar flow is characterized by smooth movement of fluid particles with ordered paths in layers with little or no lateral mixing. Turbulent flow is characterized by irregular movement of fluid particles with disordered paths. Between laminar and the turbulent flow, there is a region called transitional flow. In transitional flow, the flow switches between laminar and turbulent in a disorderly fashion. Most flows encountered in engineering practice are turbulent flows. The Reynolds number, Re, is applied to distinguish flow types. The Re for a flow from laminar type changing to the turbulent type is called the **critical Reynolds number**, Re_{cr}. For the internal flow in a circular pipe, the criteria for distinguishing different flow types are

Re ≤ 2,300 laminar flow
2,300 ≤ Re ≤ 4,000 transitional flow
Re ≥ 4,000 turbulent flow

Hydraulic Diameter

In engineering practice, pipes in flows mostly are circular shape. If noncircular pipes are involved, the pipe diameter in the Reynolds number should be replaced by the hydraulic diameter D_h defined as

$$D_h = \frac{4A_c}{P} \tag{2.35}$$

where A_c is the cross-sectional area of the flow in the pipe and P is wetted perimeter of the pipe. When there is a free surface, the wetted perimeter is the length of the walls in contact with fluid. When a full flow moving in the circular pipe, hydraulic diameter is identical to regular diameter D, i.e.,

$$D_h = \frac{4(\pi D^2/4)}{\pi D} = D$$

The relation between hydraulic diameter and hydraulic radius R_h is

$$D_h = 4R_h \qquad (2.36a)$$

and hydraulic radius is expressed as

$$R_h = \frac{A_c}{P} \qquad (2.36b)$$

The hydraulic diameters of some regular shapes are shown in Appendix A.8 Hydraulic Diameters of Regular Shapes.

2.2.4 HEAD LOSSES

As fluids flow in a pipeline, two kinds of head losses are typically encountered. One is friction head loss h_F, which is caused by the effect of the fluid viscosity. Another is local head loss h_L, which is caused by fixed components in the pipeline, such as fittings, valves, elbows, diffusers, and bends, due to the structural effect on the flow. Friction head loss and local head loss are also called major head loss and minor head loss, respectively, since the local head loss usually is much smaller than the friction head loss. These two losses form the total heat loss H_{loss} shown in Equation (2.33).

Major Head Losses

In the forced internal flow, the pressure loss due to the flow friction is determined by

$$\Delta P = f \frac{L}{D} \frac{\rho V_{avg}^2}{2} \qquad (2.37)$$

Dividing Equation (2.37) by ρg, the head loss is obtained

$$h_F = \frac{\Delta p}{\rho g} = f \frac{L}{D} \frac{V_{avg}^2}{2g} \qquad (2.38)$$

where f is the friction factor, also called the Darcy friction factor, which is a dimensionless parameter. The friction factor depends on fluid type, velocity of fluid flow, flow pattern, and roughness of pipe surface.

Colebrook Equation of Turbulent Flow

For a fully developed turbulent flow in a circular pipe, the friction factor depends on both the Reynolds number and the relative roughness. The relative roughness ε/D is the ratio of the mean height of pipe roughness to the pipe diameter. The friction factor of the turbulent flow is complicated—it cannot be derived from a theoretical analysis and is resulted from experiments or calculated by experimental formulae.

One of the formulae for calculating the friction factor is well known as the Colebrook equation obtained from experiments as shown in Equation (1.3) (see Chapter 1)

$$\frac{1}{\sqrt{f}} = -2.0 \log\left(\frac{\varepsilon/D}{3.7} + \frac{2.51}{\mathrm{Re}\sqrt{f}}\right) \tag{1.3}$$

The equation is an implicit relation in f. The determination of f using the Colebrook equation requires trial-and-error, which is not convenient. Therefore, a chart called a Moody chart based on the Colebrook equation was generated as shown in Appendix A.6 The Moody Chart. It is the chart most widely used in engineering practice for determining f in turbulent flows. For turbulent flows in noncircular pipes, the diameter D in the equation and the Moody chart is replaced by the hydraulic diameter D_h.

Haaland Equation of Turbulent Flow

The Haaland equation is an explicit formula for determining friction factor f to avoid the tedious work of trial-and-error by using the Colebrook equation. The Haaland equation is expressed as

$$\frac{1}{\sqrt{f}} = -1.8 \log\left[\frac{6.9}{\mathrm{Re}} + \left(\frac{\varepsilon/D}{3.7}\right)^{1.11}\right] \tag{2.39}$$

The result difference calculated by using the Haaland equation and the Colebrook equation is within 2%.

Minor Head Losses

Minor head loss is expressed in terms of the minor loss coefficient K_L, also called resistance coefficient. K_L is defined as

$$K_\mathrm{L} = \frac{h_L}{v^2/2g} \tag{2.40}$$

Minor head loss is, therefore, calculated by

$$h_\mathrm{L} = K_\mathrm{L}\frac{v^2}{2g} \tag{2.41}$$

K_L is determined by experiments as per the types of fittings and components. Appendix A.7(a) Minor Resistance Coefficient shows K_L of some fittings and components commonly used in engineering practice.

Total Head Loss

Once local (minor) losses and the friction (major) losses are identified, total head loss H_loss in a pipeline can be determined,

$$H_\mathrm{loss} = h_\mathrm{F} + h_\mathrm{L}$$

$$= \sum_i f_i \frac{L_i}{D_i}\frac{V_i^2}{2g} + \sum_j K_{\mathrm{L},j}\frac{V_j^2}{2g} \tag{2.42}$$

where i represents each pipe section with a constant diameter and j represents each fitting or component type. If the entire pipeline has a constant diameter, Equation (2.42) reduces to

$$H_{\text{loss}} = \left(f\frac{L}{D} + \sum_j K_{\text{L},j} \right) \frac{V^2}{2g}$$ (2.43)

where V is the average flow velocity through the entire system.

Equivalent Length L_{equiv}

Alternatively, minor head loss can be expressed in term of the equivalent length L_{equiv}, typically used in engineering practice,

$$h_{\text{L}} = K_{\text{L}}\frac{V^2}{2g} = f\frac{L_{\text{eq}}}{D}\frac{V^2}{2g}$$ (2.44)

in which

$$L_{\text{eq}} = \frac{D}{f} K_{\text{L}}$$ (2.45)

As a result, for a constant diameter pipeline, Equation (2.43) can be expressed as

$$H_{\text{loss}} = \left(L + L_{\text{eq}} \right) \frac{f}{D}\frac{V^2}{2g}$$ (2.46)

Appendix A.7(b) Equivalent Length of Valves and Fittings shows the equivalent length for steel valves and fittings commonly encountered in engineering practice.

2.2.5 PIPING SYSTEM

In engineering applications, a piping system generally is composed of a pump, straight pipes, components, and various fittings. In such an application, Equation (2.33) becomes

$$\frac{P_1}{\rho g} + \frac{1}{2}\frac{v_1^2}{g} + z_1 + H_{\text{pump}} = \frac{P_2}{\rho g} + \frac{1}{2}\frac{v_2^2}{g} + z_2 + H_{\text{loss}}$$ (2.47)

Then, required pump power is determined as

$$\dot{W} = \frac{\rho g \dot{V} H_{\text{pump}}}{\eta_{\text{pump}}}$$ (2.48a)

or

$$\dot{W} = \frac{\dot{V}\Delta P}{\eta_{\text{pump}}}$$ (2.48b)

Example 2.5

Water with density 1,000 kg/m³ and viscosity 1.02×10^{-6} m²/s is pumped in a pipeline from reservoir 1 to reservoir 2 at a steady flow 0.005 m³/s. The pipeline has a constant pipe diameter 50 mm and the length of the pipeline is 120 m with screwed fittings and components as shown. If the relative roughness of the pipe surface ε/d is 0.001 and pump efficiency is 80%, calculate (a) the pump discharge head required, (b) the minor head loss based on the equivalent length L_{eq} of fittings and components, and (c) the pump power input (kW) required to move water between the reservoirs.

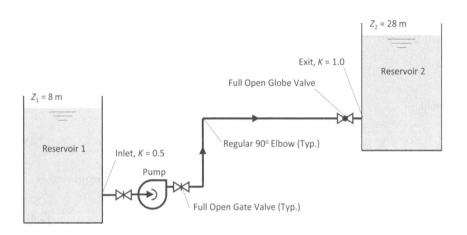

FIGURE Example 2.5

SOLUTION

Using Equation (2.47) and setting point 1 and 2 on the water surfaces of reservoir 1 and 2, respectively, the pump head in the pipeline becomes

$$H_{pump} = z_2 - z_1 + H_{loss}$$

since $P_1 = P_2$ and $V_1 = V_2 = 0$. Total head loss is determined by

$$H_{loss} = \frac{V^2}{2g}\left(\frac{fL}{d} + \sum_j K_{L,j}\right)$$

or

$$H_{loss} = \left(L + L_{eq}\right)\frac{f}{D}\frac{V^2}{2g}$$

(a) Referring to Appendix A.7(a) Minor Resistance Coefficient, the minor loss coefficients of the screwed fittings and components are

Inlet	0.5
Full open gate valve (2)	$0.17 \times 2 = 0.34$
Screwed 90° regular elbow (2)	$0.95 \times 2 = 1.9$
Full open globe valve	6.9
Exit	1.0

$$\sum_j K_{L,j} = 10.64$$

Flow velocity in the pipeline is

$$V = \frac{Q}{A} = \frac{0.005 \ \frac{m^3}{s}}{\frac{1}{4}\pi(0.05 \ m)^2} = 2.5465 \frac{m}{s}$$

Calculating the Reynolds number of the flow in the pipeline,

$$Re = \frac{VD}{v} = \frac{\left(2.5465 \ \frac{m}{s}\right)(0.05 \ m)}{1.02 \times 10^{-6} \ \frac{m^2}{s}}$$

$$= 124{,}828 > 4{,}000 \ \text{(Turbulent Flow)}$$

Referring to Appendix A.6 The Moody Chart for Re = 124,828 and $\varepsilon/d = 0.001$, the friction factor is determined to be

$$f = 0.022.$$

Substituting $\sum_i K_i$ and f into Equation (2.43), total head loss of the pipe-line becomes

$$H_{loss} = \left(\frac{0.022(120 \ m)}{0.05 \ m} + 10.64\right)\frac{\left(2.5465 \ \frac{m}{s}\right)^2}{2\left(9.81 \ \frac{m}{s^2}\right)} = 20.97 \ m$$

Therefore, the required pump discharge head is

$$H_{pump} = (28 - 8)m + H_{loss} = 20 \ m + 20.97 \ m = \textbf{40.97 m}$$

(b) Referring to Appendix A.7(b) Equivalent Length of Valves and Fittings, the total equivalent length of the minor losses is

Sharp inlet $\left(L_{eq} = \dfrac{D}{f}\, 0.5 \right)$	1.1364 m
Full open gate valve (2)	$0.5 \times 2 = 1.0$ m
Screwed 90° regular elbow (2)	$2.6 \times 2 = 5.2$ m
Full open globe valve	16.6 m
Sharp exit $\left(L_{eq} = \dfrac{D}{f}\, 1.0 \right)$	2.2727 m

$$\sum L_{eq} = 26.11 \text{ m}$$

Substituting $\sum L_{eq}$ and f into Equation (2.46), total head loss of the pipe-line is

$$H_{loss} = (120 \text{ m} + 26.11 \text{ m})\frac{0.022}{0.05 \text{ m}}\frac{\left(2.5465\,\dfrac{m}{s} \right)^2}{2\left(9.81\,\dfrac{m}{s^2} \right)} = 21.25 \text{ m}$$

It is identical to the result calculated by using the method of minor loss coefficient.

(c) Using Equation (2.48), the pump power input required to move water from reservoir 1 to reservoir 2 is

$$\dot{W} = \frac{\rho g \dot{Q} H_{pump}}{\eta_{pump}} = \left(1{,}000\,\frac{kg}{m^3} \right)\left(9.81\frac{m}{s^2} \right)\left(0.005\frac{m^3}{s} \right)(40.97 \text{ m})\,/\,0.8$$

$$= 2{,}512\frac{kg.m}{s^2}\frac{m}{s} = 2{,}512 \text{ N.}\frac{m}{s}$$

Referring to Appendix A.1 Unit Conversion, 1 N.m = 1 J and 1 J/s = 1 kW, then

$$\dot{W} = 2{,}512 \text{ N.}\frac{m}{s} = 2{,}512\,\frac{J}{s} = 2.512 \text{ kW}$$

2.3   BASICS OF HEAT TRANSFER

Heat transfer is the study of managing the rate of heat transfer, which concerns the generation, use, conversion, and exchange of thermal energy between systems. The heating and cooling of air passing through coils is the exchange of thermal energy between the fluid system flowing inside the coil and airflow. The fundamentals of heat transfer are important to the analysis and calculation of heating and cooling processes of air passing through coils.

2.3.1 HEAT TRANSFER MECHANISMS

There are three basic mechanisms of heat transfer between systems: conduction, convection, and radiation. All three heat transfer mechanisms comply with the first law of thermodynamics. Energy should be conservative in a heat transfer process,

$$\dot{Q}_{in} - \dot{Q}_{out} = \frac{dE_{sys}}{dt} \tag{2.49}$$

When heat transfer is said to be steady, i.e., system energy does not vary with time,

$$\frac{dE_{sys}}{dt} = 0$$

Thus, the heat transfer rate into and out of the system should be equal. Equation (2.49) reduces to

$$\dot{Q}_{in} = \dot{Q}_{out} = \dot{Q} \tag{2.50}$$

Conduction

The rate of steady heat conduction is described by Fourier's law as

$$\dot{Q}_{cond} = -kA\frac{dT}{dx} \tag{2.51}$$

In the equation, $\frac{dT}{dx}$ is the temperature gradient along x direction. The term is negative because the temperature decreases with increasing x. The negative sign, therefore, ensures that heat transfer in the positive x direction keeps a positive quantity. k is thermal conductivity, which is a measure of the ability of a material to conduct heat. A is the heat transfer area. The heat transfer rate is always normal to the heat transfer area.

Convection

The rate of steady heat convection is described by Newton's law as

$$\dot{Q}_{conv} = hA_s\left(T_s - T_\infty\right) \tag{2.52}$$

In the equation, h is the convection heat transfer coefficient. A_s is the surface area through which convection heat transfer takes place. T_s and T_∞ are surface temperature and surrounding temperature of the fluid sufficiently far from the surface, respectively. On the surface, fluid temperature equals surface temperature.

Radiation

The rate of steady heat radiation is described by Stefan-Boltzmann's law as

$$\dot{Q}_{rad} = \sigma A_s T_s^4 \tag{2.53}$$

In the equation, the heat transfer rate $\dot{Q}_{rad}$ is the maximum from an idealized surface called a blackbody. σ is the Stefan-Boltzmann constant and is equal to 5.670×10^{-8} $W/m^2.K^4$. The radiation emitted from a blackbody is called blackbody radiation. The radiation emitted from a real surface must be less than that from a blackbody at the same temperature. The radiation from a real surface to the ambient air that does not have a radiation effect is expressed as

$$\dot{Q}_{rad,real} = \varepsilon \sigma A_s T_s^4 \tag{2.54}$$

where ε is the emissivity of the surface. The emissivity is in the range $0 \leq \varepsilon \leq 1$. The emissivity of a blackbody is 1.

2.3.2 ONE-DIMENSIONAL HEAT CONDUCTION

One-dimensional steady heat conduction deals with the conduction in one dimension only and disregards heat conduction in the other dimensions. The application of one-dimensional heat conduction is common in engineering practice.

Heat Transfer through Single-Layer Cylinders

For one-dimensional steady heat conduction through cylindrical layers, temperature varies only along the radius direction $T(r)$. Then, Fourier's law becomes

$$\dot{Q}_{cond,cyl} = -kA \frac{dT}{dr} \tag{2.55}$$

where $A = 2\pi rL$ is heat transfer area at location r. L is the length of the cylinder. Substituting A and performing the integrations, Equation (2.55) is the form,

$$\dot{Q}_{cond,cyl} = 2\pi rL \frac{T_1 - T_2}{\ln\left(\dfrac{r_2}{r_1}\right)} \tag{2.56}$$

In calculation, Equation (2.56) is typically expressed as

$$\dot{Q}_{cond,cyl} = \frac{T_1 - T_2}{R_{cyl}} \tag{2.57}$$

where R_{cyl} is the thermal resistance of the cylindrical layer against the heat conduction,

$$R_{cyl} = \frac{\ln\left(\dfrac{r_2}{r_1}\right)}{2\pi Lk} \tag{2.58}$$

In engineering applications, fluid temperature $T_{\infty1}$ and $T_{\infty2}$ inside and outside a cylinder with the heat transfer coefficient h_1 and h_2 are usually known. The rate of heat transfer with convection on both sides, therefore, reduces to

$$\dot{Q} = \frac{T_{\infty1} - T_{\infty2}}{R_{total}} \tag{2.59}$$

The total thermal resistance R_{total} in Equation (2.59) consists of one conduction and two convection resistances,

$$R_{total} = R_{conv,1} + R_{cyl} + R_{conv,2}$$

$$= \frac{1}{(2\pi r_1 L)h_1} + \frac{\ln\left(\dfrac{r_2}{r_1}\right)}{(2\pi L)k} + \frac{1}{(2\pi r_2 L)h_2} \tag{2.60}$$

Heat Transfer through Multi-Layer Cylinders

In engineering applications, cylinders consisting of several layers of varied materials are frequently encountered. In calculation, the total thermal resistance is determined by taking a summation of thermal resistance in each layer. Figure 2.5 shows a schematic of a three-layer composite cylinder with convection on both sides. Total thermal resistance is expressed as

$$R_{total} = R_{conv,1} + R_{cyl} + R_{cy2} + R_{cy3} + R_{conv,2}$$

$$= \frac{1}{(2\pi r_1 L)h_1} + \frac{\ln\left(\dfrac{r_2}{r_1}\right)}{(2\pi L)k_1} + \frac{\ln\left(\dfrac{r_3}{r_2}\right)}{(2\pi L)k_2} + \frac{\ln\left(\dfrac{r_4}{r_3}\right)}{(2\pi L)k_3} + \frac{1}{(2\pi r_2 L)h_2} \tag{2.61}$$

Sometimes, it is convenient to express heat transfer through multiple mediums in the format of Newton's law, i.e.,

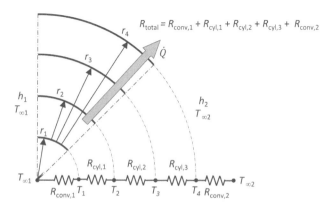

FIGURE 2.5   Heat transfer through a three-layer cylinder.

$$\dot{Q} = UA_s\Delta T \tag{2.62}$$

where U is the overall heat transfer coefficient in W/m^2.K. U is employed to calculate the rate of heat transfer from one fluid through solid surface(s) to another fluid. A comparison of Equation (2.59) and Equation (2.62) results in

$$UA_s = \frac{1}{R_{\text{total}}} \tag{2.63}$$

which means that the overall heat transfer coefficient is equal to the inverse of the total thermal resistance for a unit heat transfer area. UA_s has a unit of W/K. From Equation (2.62), it can be seen,

$$UA_s = U_i A_i = U_o A_o, \quad \text{but } U_i \neq U_o$$

When there is a condition,

$$A_i = A_o, \quad \text{then, } U_i = U_o$$

where subscriptions of i and o indicate the inside and outside of the cylinder, respectively.

Example 2.6

Steam flows in a cast iron (k = 80 W/m.K) pipeline at a temperature of 318°C. The pipe's inner and outer diameters are 5 cm and 5.5 cm, respectively. The pipeline is insulated with a 3 cm thick layer of glass wool (k = 0.05 W/m.K) and is exposed to the surrounding air at temperature of 8°C. Knowing that the convection heat transfer coefficients of inside the pipe and the surrounding are 60 W/m^2 and 18 W/m^2.K, estimate (a) the rate of heat loss from the pipeline per unit length, (b) the temperature drops across the pipe and the insulation, and (c) UA_s and the overall heat transfer coefficients U_i and U_o.

SOLUTION

The areas of the pipe's inner surface and the surface exposed to the surrounding in unit length are

$$A_1 = 2\pi r_1 L = 2\pi(0.025\text{ m})(1\text{ m}) = 0.1571\text{ m}^2$$

$$A_2 = 2\pi r_2 L = 2\pi(0.0575\text{ m})(1\text{ m}) = 0.3612\text{ m}^2$$

The thermal conductivities of cast iron and glass wool are given. The convection heat transfer coefficients of inside the pipe and surrounding are known. Then, thermal resistance of each layer is determined as

$$R_i = R_{conv,1} = \frac{1}{h_1 A_1} = \frac{1}{\left(60 \ \frac{W}{m^2.K}\right)\left(0.1571 \ m^2\right)} = 0.1061 \frac{°C}{W}$$

$$R_1 = R_{pipe} = \frac{\ln\left(r_2/r_1\right)}{(2\pi L)k_1} = \frac{\ln(2.75 \ cm/2.5 \ cm)}{2\pi (1 \ m)\left(80 \ \frac{W}{m.K}\right)} = 0.0002 \frac{°C}{W}$$

$$R_2 = R_{insul} = \frac{\ln\left(r_3/r_2\right)}{(2\pi L)k_2} = \frac{\ln(5.75 \ cm/2.75 \ cm)}{2\pi (1 \ m)\left(0.05 \ \frac{W}{m.K}\right)} = 2.35 \frac{°C}{W}$$

$$R_o = R_{conv,2} = \frac{1}{h_2 A_2} = \frac{1}{\left(18 \ \frac{W}{m^2.K}\right)\left(0.3612 \ m^2\right)} = 0.154 \frac{°C}{W}$$

Using Equation (2.61), the total resistance is

$$R_{total} = R_i + R_1 + R_2 + R_o$$
$$= (0.1061 + 0.0002 + 2.35 + 0.154)\frac{°C}{W} = 2.61\frac{°C}{W}$$

(a) The steady rate of heat loss per unit pipe length, therefore, becomes,

$$\dot{Q} = \frac{T_{\infty 1} - T_{\infty 2}}{R_{total}} = \frac{(318 - 8)°C}{2.61 \ \frac{°C}{W}} = 118.77 \ W$$

The total heat loss for the entire pipe length can be obtained by multiplying the above heat loss per unit pipe length by the entire pipe length if the length is known.

(b) The temperature drops across the pipe and insulation are

$$\Delta T_{Pipe} = \dot{Q}R_{pipe} = (118.77 \ W)\left(0.0002\frac{°C}{W}\right) = 0.0238°C$$

$$\Delta T_{insul} = \dot{Q}R_{insul} = (118.77 \ W)\left(2.35\frac{°C}{W}\right) = 279.11°C$$

(c) Using Equation (2.63), UA_s is determined to be

$$UA_s = \frac{1}{R_{total}} = \frac{1}{2.61\frac{°C}{W}} = 0.3831\frac{W}{°C} = \mathbf{0.3831\frac{W}{K}}$$

and

$$U_i = \frac{UA_s}{A_1} = \frac{0.3831\frac{W}{K}}{0.1571\,m^2} = \mathbf{2.4386\frac{W}{m^2.K}}$$

$$U_o = \frac{UA_s}{A_2} = \frac{0.3831\frac{W}{K}}{0.3612\,m^2} = \mathbf{1.0606\frac{W}{m^2.K}}$$

Heat Transfer through Finned Surfaces

From Newton's law described in Equation (2.52),

$$\dot{Q}_{conv} = hA_s\left(T_s - T_\infty\right)$$

it can be seen that the rate of heat transfer from a surface at the fixed temperature difference $T_s - T_\infty$ and the given coefficient of heat transfer h is proportional to the surface area A_s. In engineering practice, extended surfaces called fins are widely used, for instance, the finned tubes in heating and cooling coils. The attached fins usually are made of highly conductive materials, such as copper or aluminum. Finned tubes substantially increase the heat transfer effect and improve the performance of the coils. There are a variety of fin shapes in engineering applications. Some of them are shown in Figure 2.6(a).

Heat Balance of Finned Surfaces

Considering a rectangular fin at location x, the element has a length Δx and a circumference of p as shown in Figure 2.6(b). Under a steady condition, energy balance on this volume element is expressed as

$$\dot{Q}_{cond,x} = \dot{Q}_{cond,x+\Delta x} + \dot{Q}_{conv}$$

where $\dot{Q}_{cond,x}$, $\dot{Q}_{cond,\,x+\Delta x}$, and $\dot{Q}_{conv}$ are the rates of heat conduction to the element, heat conduction out of the element, and heat convection from the element, respectively. The fin temperature reduces along fin length as heat from the fin is dissipated to the surroundings. The energy balance of the entire fin, therefore, becomes

$$\dot{Q}_{fin} = \dot{Q}_{fin\ lateral\ surface} + \dot{Q}_{fin\ tip\ surface}$$

where $\dot{Q}_{fin}$, $\dot{Q}_{fin\ lateral\ surface}$, and $\dot{Q}_{fin\ tip\ surface}$ are the rates of heat transfer from the entire fin, heat conduction from the fin lateral surface, and heat convection from the fin tip surface, respectively. A temperature profile along the fin under a steady condition is shown in Figure 2.6(c)

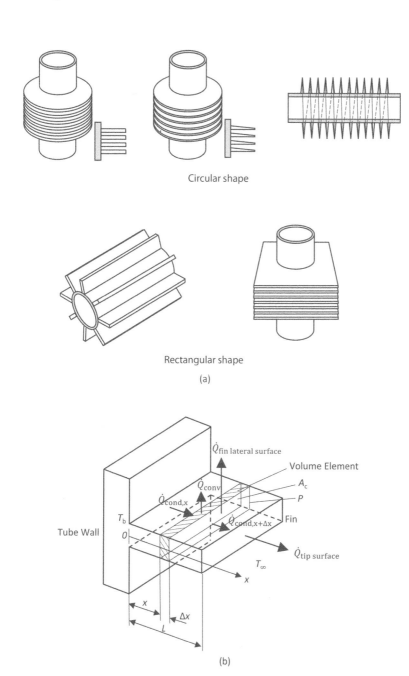

Circular shape

Rectangular shape

(a)

(b)

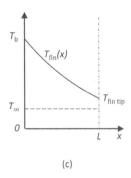

(c)

FIGURE 2.6 Fin shapes and heat balance. (a) Fin shapes. (b) Heat balance of a rectangular fin. (c) A temperature profile along a fin.

Fin Efficiency

If the temperature of an entire fin keeps the same temperature as the base temperature T_b, the rate of heat transfer from the fin is maximized. The maximized rate of heat transfer would be

$$\dot{Q}_{fin,max} = hA_{fin}\left(T_b - T_\infty\right) \tag{2.64}$$

where A_{fin} is the total surface area of the fin. In reality, however, the temperature of the fin drops along the fin and, thus, the rate of heat transfer from the fin reduces toward the fin tip as shown in Figure 2.6(c). To account for the effect of the temperature decrease on the rate of heat transfer, a fin efficiency is defined,

$$\eta_{fin} = \frac{\dot{Q}_{fin}}{\dot{Q}_{fin,max}} = \frac{\text{Actual heat transfer rate from the fin}}{\text{Ideal heat transfer rate from the fin}}$$
$$\text{if the entire fin has the base temperature}$$

The rate of heat transfer from an actual fin is

$$\dot{Q}_{fin} = \eta_{fin}\dot{Q}_{fin,\,max} = \eta_{fin}hA_{fin}\left(T_b - T_\infty\right) \tag{2.65}$$

When fin efficiency is known, the rate of heat transfer from an actual fin can be decided by using Equation (2.65). In some applications, the weight of finned coils is a major concern, such as in aircrafts and space shuttles. Fins with higher efficiency and less weight are always preferable.

Fin Effectiveness

The purpose of fin application is to enhance the rate of heat transfer. The performance of tubes or plates with fins is evaluated by the rate of heat transfer enhanced with fins compared to that without fins. The rate is defined as fin effectiveness, ε_{fin},

$$\varepsilon_{\text{fin}} = \frac{\text{Actual heat transfer rate from the fin of base area } A_b}{\text{Heat transfer rate from the surface area } A_b \text{ without fins}}$$

(2.66)

$$= \frac{\dot{Q}_{\text{fin}}}{\dot{Q}_{\text{no fin}}} = \frac{\dot{Q}_{\text{fin}}}{hA_b(T_b - T_\infty)}$$

where A_b represents the area covered by the fin root at the base surface and $\dot{Q}_{\text{no fin}}$ represents the rate of heat transfer from this area without fin attachment. From the view of heat transfer enhancement, ε_{fin} should be larger than 1. The bigger the ε_{fin} is, the better the result with fins has, and the higher the rate of heat transfer with fins is obtained.

Fin efficiency and fin effectiveness are two different descriptions. But they have a relationship. Using Equations (2.65) and (2.66), their relationship is

$$\varepsilon_{\text{fin}} = \frac{\dot{Q}_{\text{fin}}}{\dot{Q}_{\text{no fin}}} = \frac{\dot{Q}_{\text{fin}}}{hA_b(T_b - T_\infty)}$$

(2.67)

$$= \frac{\eta_{\text{fin}} h A_{\text{fin}}(T_b - T_\infty)}{hA_b(T_b - T_\infty)} = \eta_{\text{fin}} \frac{A_{\text{fin}}}{A_b}$$

Fin effectiveness can be determined conveniently when fin efficiency is known, or vice versa.

Example 2.7

Steam in a heating system flows through a vertical tube. The tube has an outer diameter of 40 mm and maintains a constant surface temperature of 120°C. Circular aluminum alloy fins ($k = 186$ W/m·K) with an outer diameter of 70 mm and a thickness of 1 mm are attached to the tube as shown. Space between the

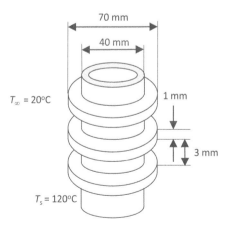

FIGURE Example 2.7

fins is 3 mm, thus, in total, there are 250 fins per meter length of the tube. If heat is transferred to the surrounding air at $T_\infty = 20°C$ with a heat transfer coefficient $h = 40$ W/m²·K and knowing fin efficiency is 0.94, determine *(a)* the increase of the heat transfer rate (W) from the tube per meter length as a result of adding fins and *(b)* the increase percentage (%) and times of the heat transfer rate from the finned tube.

SOLUTION

(a) Thermal conductivities of the fin material, the heat transfer coefficient of the surrounding, and fin efficiency are given. The rate of heat transfer from the tube per meter length without fins is

$$A_{nofin} = \pi D_1 L = \pi (0.04 \text{ m})(1 \text{ m}) = 0.1257 \text{ m}^2$$

$$\dot{Q}_{nofin} = hA_{nofin}(T_b - T_\infty)$$

$$= (40 \text{ W/m}^2°C)(0.1257 \text{ m}^2)(120 - 20)°C = 502.8 \text{ W}$$

The surface of a single fin is

$$A_{fin} = 2\pi (r_2^2 - r_1^2) + 2\pi r_2 t = 2\pi (0.035^2 - 0.02^2)\text{m}^2 + 2\pi (0.035)(0.001)\text{m}^2$$

$$= 0.0052 \text{ m}^2 + 0.0002 \text{ m}^2 = 0.0054 \text{ m}^2$$

Then, the rate of heat transfer from a single fin becomes

$$\dot{Q}_{fin} = \eta_{fin}\dot{Q}_{fin,max} = \eta_{fin}hA_{fin}(T_b - T_\infty)$$

$$= 0.94\left(40\frac{W}{m^2.K}\right)(0.0054 \text{ m}^2)(120 - 20)°C = 20.3 \text{ W}$$

The rate of heat transfer from a single unfinned part of the tube is determined to be

$$A_{unfin} = \pi D_1 s = \pi (0.04 \text{ m})(0.003 \text{ m}) = 0.00038 \text{ m}^2$$

$$\dot{Q}_{unfin} = hA_{fin}(T_b - T_\infty) = \left(40\frac{W}{m^2.K}\right)(0.00038 \text{ m}^2)(120 - 20)°C = 1.52 \text{ W}$$

There are 250 fins and, thus, 250 spaces between fins per meter length of the tube. The rate of total heat transfer from the finned tube is

$$\dot{Q}_{total,fin} = n(\dot{Q}_{fin} + \dot{Q}_{unfin})$$

$$= 250(20.3 + 1.52)W = 5,455 \text{ W}$$

Therefore, the increase of the rate of heat transfer rate from the finned tube compared to the tube without fins per meter length becomes

$$\dot{Q}_{increase} = \dot{Q}_{total,fin} - \dot{Q}_{nofin}$$
$$= (5,455 - 502.8)W = \mathbf{4,952.2\ W}$$

(b) The increased percentage (%) of the heat transfer rate from the finned tube per meter length is determined to be

$$\Delta\dot{Q} = \frac{\dot{Q}_{total,fin} - \dot{Q}_{nofin}}{\dot{Q}_{nofin}}$$
$$= \frac{(5,455 - 502.8)W}{502.8\ W} = 9.84.9 = \mathbf{984.9\%}$$

$$\Delta = \frac{5,455\ W}{502.8\ W} = \mathbf{10.85\ times}$$

2.3.3 FORCED CONVECTIVE HEAT TRANSFER

Forced convective heat transfer can be classified as internal or external. Internal forced convection occurs as fluid flows inside a conduit and external forced convection occurs as fluid flows outside a conduit. For instance, in heating and cooling of air passing through coils, fluid flowing inside the coil is forced internal convection and air passing through the coil is forced external convection.

Internal Forced Convection

Referring to Equations (2.3b) and (2.31b), the rate of heat transfer for steady fluid flow in a tube is

$$\dot{Q} = \dot{m}\Delta h = m(h_e - h_i) = mc_p(T_e - T_i) \qquad (2.68)$$

where T_i and T_e are the temperatures of fluid entering and exiting the tube, respectively. The rate of heat transferred through the tube surface is expressed by Newton's law,

$$\dot{Q} = hA_s(T_s - T_m)_{avg} = hA_s\Delta T_{avg} \qquad (2.69)$$

where h is the average convection heat transfer coefficient, A_s is the heat transfer surface area, T_s is the tube surface temperature, and T_m is the mean temperature of the fluid. In the case of a constant T_s, ΔT_{avg} can be estimated by the log mean temperature difference (LMTD) ΔT_{lm}, i.e.,

$$\Delta T_{\text{avg}} = \Delta T_{\text{lm}} = \frac{(T_s - T_e) - (T_s - T_i)}{\ln\left[(T_s - T_e)/(T_s - T_i)\right]} \tag{2.70}$$

$$= \frac{\Delta T_e - \Delta T_i}{\ln(\Delta T_e)/(\Delta T_i)}$$

where $\Delta T_i = T_s - T_i$ and $\Delta T_e = T_s - T_e$ are the temperature differences between the surface and the fluid at the inlet and the exit of the tube, respectively.

Example 2.8

Water at a mass rate of 0.5 kg/s and a temperature of 10°C enters a copper tube of 3.0 cm internal diameter in a heat exchanger. The water is heated to 120°C by steam condensing outside the tube at temperature of 130°C. If knowing the average heat transfer coefficient is 850 W/m².K and the specific heat of the water at the bulk mean temperature is 4,187 J/kg.K, determine (a) the heat transfer rate of the heat exchanger (kW), (b) the LMTD, (c) the length L (m) of the tube required to heat the water to the desired temperature, and (d) the bundle length l (m) if the heat exchanger is in 16-pass (8 U-bend) design.

SOLUTION

(a) Using Equation (2.68), the rate of heat transfer of the heat exchanger is

$$\dot{Q} = \dot{m} c_p \left(T_e - T_i\right)$$

$$= \left(0.5 \text{ kg}/\text{s}\right)\left(4.187 \frac{\text{kJ}}{\text{kg.K}}\right)(120 - 10)\,°\text{C} = \textbf{230.29 kW}$$

(b) The temperature difference between the surface and fluid at the inlet and exit of the tube is

$$\Delta T_e = \left(T_s - T_e\right) = 130°\text{C} - 120°\text{C} = 10°\text{C}$$

$$\Delta T_i = \left(T_s - T_i\right) = 130°\text{C} - 10°\text{C} = 120°\text{C}$$

Using Equation (2.70), The LMTD is obtained,

$$\Delta T_{\text{lm}} = \frac{\Delta T_e - \Delta T_i}{\ln(\Delta T_e)/(\Delta T_i)} = \frac{(10 - 120)\,°\text{C}}{\ln\left(\dfrac{10}{120}\right)} = \textbf{44.27°C}$$

(c) Using Equation (2.69), the heat transfer surface area is determined to be

$$A_s = \frac{\dot{Q}}{h \Delta T_{\text{lm}}} = \frac{230.29 \text{ kW}}{\left(0.85 \dfrac{\text{kW}}{\text{m}^2}.\text{K}\right)(44.27°\text{C})} = 6.1199 \text{ m}^2$$

Then using $A_s = \pi D L$, the required tube length is

$$L = \frac{A_s}{\pi D} = \frac{6.1199 \text{ m}^2}{\pi\,(0.03 \text{ m})} = \textbf{64.93 m}$$

(d) The bundle length l in the heat exchange with 16-pass is determined as

$$l = \frac{L}{n} = \frac{64.93 \text{ m}}{16} = \textbf{4.06 m}$$

External Forced Convection

Crossflow over a tube bank is commonly encountered in heat transfer equipment, which is an external forced convection through tubes. Figure 2.7 shows two typical tube banks. One is in-line and another is staggered along the flow direction. The arrangement of the tubes in tube banks is identified by the transverse pitch S_T, the longitudinal pitch S_L, and the diagonal pitch S_D between tube centers. The diagonal pitch S_D in the staggered tube bank is decided by the transverse pitch S_T and the longitudinal pitch S_L as

$$S_D = \sqrt{S_L^2 + (S_T/2)^2}$$

• *Maximum Velocity*

As fluid enters a tube bank, the flow area reduces from $A_1 = S_T L$ to $A_T = (S_T - D)L$ between the tubes and, thus, flow velocity increases. In staggered arrangement, the velocity may increase further in the diagonal region if the tube rows are very close to each other. In the tube bank, the outer tube diameter D is taken as the characteristic length and the flow characteristics are dominated by the maximum velocity V_{max} that

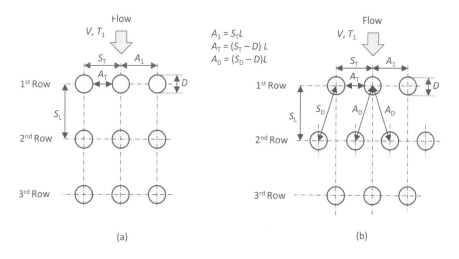

(a)   (b)

FIGURE 2.7   Arrangement of tubes in tube banks. (a) In-line. (b) Staggered.

occurs at the minimum flow area between the tubes. The Reynolds number, there-fore, is determined on the basis of maximum velocity V_{max} as

$$\text{Re}_D = \frac{V_{max}D}{v} = \frac{\rho V_{max}D}{\mu} \tag{2.71}$$

The maximum velocity can be decided from the mass conservative of incompress-ible steady flow.

For in-line arrangement, the conservation of mass is

$$\rho VA_1 = \rho V_{max} A_T$$

or

$$VS_T = V_{max} \left(S_T - D\right)$$

Then the maximum velocity is determined to be

$$V_{max} = \frac{S_T}{S_T - D} V \tag{2.72a}$$

For staggered arrangement, the fluid flowing through area A_1 passes through area A_T and then through area $2A_D$ as the fluid approaches the pipes in the next row. If $2A_D > A_T$, Equation (2.72a) is still valid for staggered tube banks. Otherwise, if $2A_D < A_T$ or $2(S_D - D) < (S_T - D)$, the maximum velocity occurs at the diagonal cross sections, which is

$$V_{max} = \frac{S_T}{2(S_D - D)} V \tag{2.72b}$$

The nature of flow around a tube in the first row resembles flow over a single tube, especially when the tubes are not too close to each other. Each tube in a tube bank that consists of a single transverse row can be treated as a single tube in crossflow. The nature of flow around a tube in the second and subsequent rows, however, is different because of wakes formed and the turbulence caused by the tubes upstream. The level of turbulence and thus the heat transfer coefficient increases with row number due to the combined effects of upstream rows.

• *Heat Transfer Coefficient*

The heat transfer coefficient h can be determined from the Nusselt number Nu. The Nusselt number characterized by the tube outside diameter D is defined as

$$Nu_D = \frac{hD}{k} \tag{2.73}$$

The average heat transfer coefficient for the entire tube bank depends on the number of tube rows along the flow as well as the arrangement and the size of the tubes. Flow through tube banks is complex to be treated analytically. The Nu_D correlations are based on experimental data. Once the average heat transfer coefficient based on the Nusselt number for the entire tube bank is determined, the rate of heat transfer can be decided from Newton's law of heat convention using LMTD ΔT_{lm} from Equation (2.69). The exit temperature of the fluid exiting the tube bank becomes

$$T_e = T_s - (T_s - T_i)\exp\left(-\frac{A_s h}{\dot{m}c_p}\right) \quad (2.74)$$

In the equation, $A_s = N\pi DL$ is the heat transfer surface area, $\dot{m} = \rho V(N_T S_T L)$ is the mass flow rate of the fluid, N is the total number of tubes in the bank, which is the product of N_T (number of tubes in the transverse plane) and N_L (number of rows in the flow direction), L is the length of the tubes, and V is the velocity of the fluid in front of the tube bank. The rate of heat transfer rate through the tube bank is decided to be

$$\dot{Q} = hA_s\Delta T_{\text{lm}} = \dot{m}c_p\left(T_e - T_i\right) \quad (2.75)$$

- *Pressure Drop*

The pressure drop ΔP of the fluid between the inlet and the exit of the tube bank is from the tubes to against fluid flow over them, which is determined by

$$\Delta P = N_L f\chi \frac{\rho V_{\text{max}}^2}{2} \quad (2.76)$$

where N_L is the rows in the flow direction, f is the friction factor, and χ is the correction factor. Both f and χ depend on the Reynolds number in the maximum velocity.

- *Required Power*

When the pressure drop calculated from Equation (2.75) is known, the power required to overcome the friction resistance in the tube bank can be determined by

$$\dot{W} = \dot{V}\Delta P = \frac{\dot{m}}{\rho}\Delta P \quad (2.77)$$

where $\dot{V} = V(N_T S_T L)$ is the volume flow rate of the fluid through the tube bank. The required power is proportional to the pressure drop. Therefore, using a finned tube bank to enhance the rate of heat transfer should be weighed to increase power consumption since the finned tubes will enlarge the pressure drop as fluid passes through the tube bank.

Example 2.9

Air is heated by passing through a tube bank of 4 m long at a mean velocity of 5.5 m/s. Steam flowing inside the tubes is used to heat the air and condenses at 100°C. Air approaches the tube bank in a normal direction at temperature of 20°C and a pressure of 1 atm. The outer diameter of the tubes is 1.6 cm. The tubes are in a staggered arrangement with longitudinal and transverse pitches of

$S_L = S_T = 0.04$ m. The bank has 20 rows in the flow direction and 13 tubes in each row. For such a tube bank, the friction factor f and correction factor χ are 0.32 and 1, respectively. If air properties are evaluated by a mean temperature of 35°C at 1 atm ($k = 0.02625$ W/m.K) and knowing Nu_D is 73.29 from the experiment, determine (a) the maximum velocity in the tube bank (m/s), (b) the rate of heat transfer (W) through the tube bank, (c) the pressure drop (Pa) across the tube bank, and (d) the condensation rate of the steam (kg/s) inside the tubes.

SOLUTION

Referring to Appendix A.3 Specific Heat of Air at mean temperature of 35°C, the following properties are found,

$$\rho = 1.1457 \frac{kg}{m^3} \quad c_p = 1.006 \frac{kJ}{kg.K}$$

The density of air at the inlet temperature of 20°C is

$$\rho_i = 1.2043 \frac{kg}{m^3}.$$

Referring to Appendix A.5 Properties of Steam and Compressed Water, the enthalpy of steam condensation at temperature 100°C is

$$h_{fg\,@\,Ts} = 100°C = 2,256.4 \frac{kJ}{kg}$$

(a) Knowing $D = 0.016$ m, $S_L = S_T = 0.04$ m, and $V = 5.5$ m/s, the maximum velocity is determined from Equation (2.72a) since $S_D > (S_T + D)/2$,

$$V_{max} = \frac{S_T}{S_T - D} V$$

$$= \frac{0.04 \text{ m}}{0.04 \ m - 0.016 \ m} \left(5.5 \frac{m}{s} \right) = \textbf{9.1667} \frac{\textbf{m}}{\textbf{s}}$$

(b) The Nusselt number is given, $Nu_D = 73.29$. The heat transfer coefficient can be found by Equation (2.73),

$$h = \frac{Nu_D k}{D} = \frac{73.29 \left(0.02625 \frac{W}{m.°C} \right)}{0.016 \text{ m}} = 120.24 \frac{W}{m^2.°C}$$

The total number of tubes is $N = N_L \times N_T = 20 \times 13 = 260$. For the given tube length $L = 4$ m, the heat transfer surface area and the mass flow rate of air (evaluated at the inlet) are determined as

$$A_s = N\pi DL = 260\pi (0.016 \text{ m})(4 \text{ m}) = 52.28 \text{ m}^2$$

$$\dot{m} = \dot{m}_i = \rho_i V (N_T S_L L)$$

$$= \left(1.2043 \frac{kg}{m^3}\right)\left(5.5 \frac{m}{s}\right)13(0.04\ m)(4\ m) = 13.78 \frac{kg}{s}$$

Using Equation (2.74), the fluid exit temperature is

$$T_e = T_s - (T_s - T_i)\exp\left(-\frac{A_s h}{\dot{m}c_p}\right)$$

$$= 100°C - (100°C - 20°C)\exp\left[-\frac{(52.28\ m^2)\left(120.24 \dfrac{W}{m^2.°C}\right)}{\left(13.78 \dfrac{kg}{s}\right)\left(1,006 \dfrac{J}{kg.°C}\right)}\right]$$

$$= 100°C - (80°C)\exp(-0.4535) = 49.17°C$$

Using Equation (2.70), the LMTD is determined to be

$$\Delta T_{lm} = \frac{(T_s - T_e) - (T_s - T_i)}{\ln\left[(T_s - T_e)/(T_s - T_i)\right]}$$

$$= \frac{(100 - 49.17)°C - (100 - 20)°C}{\ln\left[(100 - 49.17)°C / (100 - 20)\right]°C} = 64.32°C$$

Using Equation (2.75), the rate of heat transfer, therefore, is

$$\dot{Q} = hA_s\Delta T_{lm} = \left(120.24\frac{W}{m^2.°C}\right)(52.28\ m^2)(64.32°C)$$

$$= \mathbf{404,325\ W}$$

(c) Knowing $f = 0.32$ and $\chi = 1$, the pressure drop across the tube bank by using Equation (2.76) is found,

$$\Delta P = N_L f \chi \frac{\rho V_{max}^2}{2}$$

$$= 20(0.32)(1)\frac{\left(1.1457 \dfrac{kg}{m^3}\right)\left(9.1667 \dfrac{m}{s}\right)^2}{2}\left(\frac{1\ N}{1\ kg.\dfrac{m}{s^2}}\right) = \mathbf{308.07\ Pa}$$

(d) Using Equation (2.1),

$$\dot{Q} = \dot{m}_{cond} h_{fg@100°c}$$

the rate of condensation of steam is determined to be

$$\dot{m}_{cond} = \frac{\dot{Q}}{h_{fg@100°C}}$$

$$= \frac{404.325 \text{ kW}}{2,256.4 \dfrac{\text{kJ}}{\text{kg}}} = \textbf{0.1792} \ \frac{\textbf{kg}}{\textbf{s}}$$

2.3.4 THERMAL RADIATION

Radiation is a process of energy transfer from a source and travels through space at the speed of light. Unlike conduction and convection, heat transfer by radiation does not require the presence of an intervening medium. Anything at a temperature above absolute zero can emit thermal radiation.

Blackbody Radiation

A blackbody is an ideal body that can absorb all incident radiation falling on it. The radiation energy emitted by a blackbody per unit time and per unit surface area is determined by Stefan-Boltzmann's law, which is

$$E_b(T) = \sigma T^4 \tag{2.78a}$$

or in rate,

$$\dot{Q}_b(T) = \sigma A T^4 \tag{2.78b}$$

where $\sigma = 5.670 \times 10^{-8}$ W/m^2.k^4 is the Stefan-Boltzmann constant and T is the absolute temperature of the body surface in K, respectively. Radiation emitted by real surfaces (gray) is less than radiation emitted by a blackbody at the same temperature expressed as

$$E(T) = \varepsilon \sigma T^4 \tag{2.79a}$$

or in rate,

$$\dot{Q}(T) = \varepsilon \sigma A T^4 \tag{2.79b}$$

where ε is the emissivity of the surface. The value of the property emissivity range is $1 \leq \varepsilon \leq 1$. Emissivity is a measure of how closely a real surface approximates a blackbody ($\varepsilon = 1$).

Irradiation

Irradiation or incident radiation G is the amount of radiation falling on a surface. The unit of irradiation is W/m^2, which represents the rate at which radiation energy is incident on a surface per unit area of the surface.

Radiative Properties

When incident radiation G strikes on a semitransparent surface, part of it is absorbed, part of G is reflected, and the remaining part, if any, is transmitted, as illustrated in Figure 2.8.

The fraction of irradiation absorbed by the surface is called the absorptivity, the fraction reflected by the surface is called the reflectivity, and the fraction transmitted is called the transmissivity. They are

$$\text{Absorptivity} \quad \alpha = \frac{\text{Absorbed radiation}}{\text{Incident radiation}} = \frac{G_{abs}}{G} \quad 0 \le \alpha \le 1 \quad (2.80a)$$

$$\text{Reflectivity} \quad \rho = \frac{\text{Reflected radiation}}{\text{Incident radiation}} = \frac{G_{ref}}{G} \quad 0 \le \rho \le 1 \quad (2.80b)$$

$$\text{Transmissivity} \quad \tau = \frac{\text{Transmitted radiation}}{\text{Incident radiation}} = \frac{G_{tr}}{G} \quad 0 \le \tau \le 1 \quad (2.80c)$$

where G_{abs}, G_{ref}, and G_{tr} are the absorbed, reflected, and transmitted portions of the irradiation on the surface, respectively. The first law of thermodynamics requires that the sum of the absorbed, reflected, and transmitted radiation be equal to the incident radiation which is,

$$G_{abs} + G_{ref} + G_{tr} = 1 \quad (2.81a)$$

Dividing each term of this relation by G yields

$$\alpha + \rho + \tau = 1 \quad (2.81b)$$

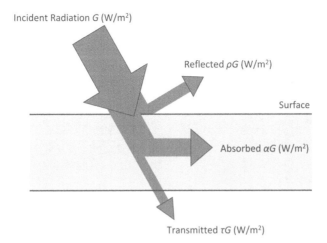

Incident Radiation G (W/m²)

Reflected ρG (W/m²)

Surface

Absorbed αG (W/m²)

Transmitted τG (W/m²)

FIGURE 2.8 Absorption, reflection, and transmission.

For idealized blackbodies, which are perfect absorbers, $\rho = 0$ and $\tau = 0$, then Equation (2.81b) reduces to $\alpha = 1$. For opaque surfaces, such as most solids and liquids, Equation (2.81b) becomes

$$\alpha + \rho = 1 \qquad\qquad (2.81c)$$

For most gases, reflectance is absent, i.e., $\rho = 0$. Then Equation (2.81b) becomes

$$\alpha + \tau = 1 \qquad\qquad (2.81d)$$

Equations (2.81b) and (2.81c) are important since knowing one of the properties among α, ρ, and τ implies the knowledge of the other properties.

3 The Psychrometric Chart

Psychometrics is a branch of science that studies the physical and thermodynamic properties of moist air. A psychrometric chart graphically represents the physical and thermodynamic properties of moist air, the relationship between the properties, and processes of the property change of moist air. The psychrometric chart is an important and useful tool to identify properties and analyze the heating and cooling processes of air passing through coils.

3.1 PSYCHROMETRIC PROPERTIES OF AIR

Atmospheric air is a pure substance. It is a mixture of moisture, air, and other substances. Air that does not contain moisture is called dry air. The state of air can be completely specified by two independent intensive properties such as temperature, specific volume, humidity, enthalpy, and so on. The remaining properties of the state can be identified by using the ideal gas equation of state, other relations, or the psychrometric chart.

3.1.1 SPECIFIC AND RELATIVE HUMIDITY

Humidity of air is defined as the amount of water vapor in the air, which can be treated as an ideal gas. The humidity can be specified in two ways: specific humidity and relative humidity. If the humidity is defined as the mass of water vapor m_v presented in a unit mass of dry air m_a, it is called specific humidity ω or absolute humidity,

$$\omega - \frac{m_v}{m_a} \tag{3.1}$$

The unit of the specific humidity ω is kg H_2O/kg dry air. By using the ideal gas equation of state, the specific humidity ω can also be expressed as

$$\omega = \frac{m_v}{m_a} = \frac{P_v/R_v T}{P_a/R_a T} = \frac{P_v/R_v}{P_a/R_a} = 0.622 \frac{P_v}{P_a} \tag{3.2a}$$

or

$$\omega = \frac{0.622 P_v}{P - P_v} \tag{3.2b}$$

where P is the total pressure of the air, which is the sum of the partial pressure of dry air P_a and water vapor P_v, i.e.,

$$P = P_a + P_v \tag{3.3}$$

DOI: 10.1201/9781003289326-3

The comfort level of humans, however, depends more on relative humidity (RH) ϕ. The RH is defined as the ratio of the mass of water vapor m_v in the air to the maximum mass of moisture m_g the air can hold at the same temperature,

$$\phi = \frac{m_v}{m_g} = \frac{P_v / R_v T}{P_g / R_v T} = \frac{P_v}{P_g} \qquad (3.4)$$

where $P_g = P_{sat@T}$

Combining Equations (3.2b) and (3.4), the relation of the specific humidity and relative humidity is obtained as

$$\phi = \frac{\omega P}{(0.622 + \omega) P_g} \qquad (3.5a)$$

or

$$\omega = \frac{0.622 \phi P_g}{P - \phi P_g} \qquad (3.5b)$$

The RH ranges from 0 for dry air to 1 for saturated air, or from 0% to 100% in percentage. The amount of moisture that air can hold depends on the air temperature. Therefore, even when the specific humidity of the air remains constant under a given process, the RH of the air may vary.

3.1.2 SPECIFIC VOLUME

Specific volume of air v is defined as the volume per unit mass of air. The unit of the specific volume of atmospheric air is m^3/kg dry air. The specific volume of air changes with variation of atmospheric pressure, temperature, and humidity, which can be determined by the ideal gas equation of state, Equation (1.1) shown in Section 1.2 (see Chapter 1),

$$Pv = RT$$

i.e.,

$$v = \frac{RT}{P} \qquad (3.6)$$

The density of air ρ is inverse to the specific volume of air v and described by Equation (2.28b) in Section 2.2.1(see Chapter 2),

$$\rho = \frac{1}{v}$$

3.1.3 Dry-Bulb Temperature

The ordinary temperature of atmospheric air is referred to as dry-bulb temperature T_{db}. It is called "dry-bulb" because the air temperature is measured by a thermometer not affected by the moisture of the air. The dry-bulb temperature is also the ambient temperature commonly used in engineering practices.

3.1.4 Wet-Bulb Temperature

Wet-bulb temperature T_{wb} is the lowest temperature to which air can be cooled by evaporating water into the air at a constant pressure. The wet-bulb temperature is measured by a thermometer wrapped in wet muslin. Therefore, the measurement of the wet-bulb temperature is affected by the moisture content in the air. The wet-bulb temperature is always lower than the dry-bulb temperature due to evaporative cooling. The difference between the wet-bulb temperature and dry-bulb temperature is a measure of the humidity of the air. At 100% relative humidity, the wet-bulb temperature is equal to the dry-bulb temperature.

The dry-bulb temperature T_{db} and wet-bulb temperature T_{wb} can be measured simultaneously by a device called sling psychrometer, which mounts both T_{db} and T_{wb} thermometers on the same frame. The specific humidity of air, therefore, can be decided by

$$\omega_1 = \frac{c_p(T_2 - T_1) + \omega_2 h_{fg2}}{h_{g1} - h_{f2}} \qquad (3.7a)$$

where

$$\omega_2 = \frac{0.622\, P_{g2}}{P_2 - P_{g2}} \qquad (3.7b)$$

which is derived from Equation (3.5b). In the equations, subscripts 1 and 2 correspond to T_{db} and T_{wb}, respectively.

3.1.5 Dew-Point Temperature

Dew-point temperature T_{dp} is defined as the temperature at which condensation of water vapor in the air begins when the air is cooled at a constant pressure. T_{dp} is the saturation temperature of water corresponding to the vapor pressure,

$$T_{dp} = T_{sat@\ P_v} \qquad (3.8)$$

At a dew point, air is saturated and its relative humidity ϕ is 100%. The dry-bulb temperature T_{db}, wet-bulb temperature T_{wb}, and dew-point temperature T_{dp} at the dew point are identical. Any further drop in temperature results in condensation of some of the moisture from the air.

3.1.6 Specific Enthalpy

Specific enthalpy of air h is defined as the enthalpy per unit mass of dry air. Since atmospheric air is a mixture of dry air and moisture, the total enthalpy of atmospheric air H is a summation of the enthalpies of the dry air and the water vapor, i.e.,

$$H = H_a + H_v \tag{3.9a}$$

or

$$H = m_a h_a + m_v h_v \tag{3.9b}$$

Dividing Equation (3.9b) by m_a,

$$h = \frac{H}{m_a} = h_a + \frac{m_v}{m_a} h_v = h_a + \omega h_v \tag{3.10a}$$

Since $h_v \cong h_g$, Equation (3.10a) then becomes

$$h = h_a + \omega h_g \tag{3.10b}$$

Thus, the total specific enthalpy of air h is the summation of specific sensible heat and specific evaporated heat of the air. The unit of the specific heat of air h is kJ/kg dry air. The specific evaporated heat represents specific latent heat of the water evaporated in the air. The enthalpies of total specific, specific sensible, and specific latent are very useful in the analysis of cooling and heating processes of air passing through coils.

Example 3.1

If the air in a room at 1 atm (101.325 kPa) pressure has 30°C T_{db} and 20°C T_{wb}, respectively, determine (a) the specific humidity ω, (b) the relative humidity ϕ, (c) the enthalpy, and (d) the dew-point temperature T_{dp} of the air at the given condition.

SOLUTION

The state of the air in the room is specified by the given pressure and temperature. Referring to Appendix A.5 Properties of Steam and Compressed Water, the saturated pressures of water are,

$$P_{g2} = P_{s@20°C} = 2.3393 \text{ kPa}$$

$$P_{g1} = P_{s@30°C} = 4.2470 \text{ kPa}$$

and the evaporation enthalpy is

$$h_{fg} = h_{fg@20°C} = 2,453.5 \ \frac{\text{kJ}}{\text{kg}}$$

The enthalpies of vapor and water at 30°C T_{db} and 20°C T_{wb} become

$$h_{g1} = 2,555.5\frac{kJ}{kg}$$

$$h_{f2} = 83.91\frac{kJ}{kg}$$

Referring to Appendix A.3 Specific Heat of Air, the specific heat of air at the room condition is

$$c_p = 1.006\frac{kJ}{kg.°C}$$

(a) Using Equation (3.7b),

$$\omega_2 = \frac{0.622\, P_{g2}}{P_2 - P_{g2}}$$

$$= \frac{0.622(2.3393\text{ kPa})}{(101.325 - 2.3393)\text{kPa}} = 0.0147\frac{kg\ H_2O}{kg\ dry\ air}$$

Using Equation (3.7a). the specific humidity of the air is found,

$$\omega_1 = \frac{c_p(T_2 - T_1) + \omega_2 h_{fg2}}{h_{g1} - h_{f2}}$$

$$= \frac{\left(1.006\frac{kJ}{kg.°C}\right)(20 - 30)°C + \left(0.0147\frac{kg\ H_2O}{kg\ dry\ air}\right)\left(2,453.5\frac{kJ}{kg}\right)}{(2,555.5 - 83.91)\frac{kJ}{kg}}$$

$$= 0.0105\frac{kg\ H_2O}{kg\ dry\ air}$$

(b) Using Equation (3.5a), the relative humidity ϕ_1 is decided as

$$\phi_1 = \frac{\omega_1 P_2}{(0.622 + \omega_1)P_{g1}}$$

$$= \frac{0.0105\,(101.325\text{ kPa})}{(0.622 + 0.0105)(4.2470)\text{ kPa}} = 0.3961\text{ or }39.61\%$$

(c) Using Equations (3.10b) and (2.31b) (see Chapter 2), the specific enthalpy of air is determined to be

$$h_1 = h_{a1} + \omega_1 h_{g1} = c_p T_1 + \omega_1 h_{g1}$$

$$= \left(1.006 \ \frac{kJ}{kg.°C}\right)(30°C) + (0.0105)\left(2,555.5 \ \frac{kJ}{kg}\right)$$

$$= 57.01 \ \frac{kg}{kg \ dry \ air}$$

(d) Using Equation (3.4), the vapor pressure of the air is obtained,

$$P_{v1} = \phi_1 P_{g1} = \phi_1 P_{sat\,@\,30°C}$$
$$= 0.3961(4.2470 \ kPa) = 1.6822 \ kPa$$

Referring to Appendix A.5 Properties of Steam and Compressed Water, the dew-point temperature of the air is

$$T_{dp} = T_{sat\,@\,1.6818 \ kPa} = \mathbf{14.69°C}$$

In engineering practice, the psychrometric chart as shown below (see Chapter 1) is commonly used to conveniently determine the states and properties of air.

3.2   OVERVIEW OF THE PSYCHROMETRIC CHART

A state of atmospheric air at a specified pressure is completely identified by two independent intensive properties. By using the psychrometric chart, other properties of the state can be determined. The psychrometric chart shown in Section 1.2 (see Chapter 1) is presented below to be used for studying the processes of air at a pressure of 1 atm (101.325 kPa). The chart is also attached in Appendix A.4.

In the chart, the horizontal axis represents the dry-bulb temperature T_{db}. The vertical axis on the right end represents the specific humidity ω. The curved line on the left side is called the saturation line. All points on the curve represent the saturated air states, which is also the curve of 100% RH. Other constant relative humidity ϕ curves in the chart have the same general shapes as that of the saturation line. The lines of constant wet-bulb temperature T_{wb} have a downhill appearance to the right. The lines of constant specific volume v have a similar downhill appearance to the right as the lines of the constant wet-bulb temperature, but they are steeper. The lines of constant enthalpy h lie very nearly parallel to the lines of constant wet-bulb temperature.

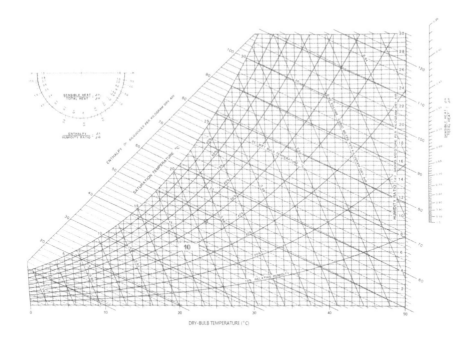

DRY-BULB TEMPERATURE (°C)

FIGURE 1.6   Psychrometric chart at a pressure of 1 atm (101.325 kPa).

Example 3.2

A room at 1 atm pressure contains 30 kg of air mass at 35°C T_{db} and 30% RH. Using the psychrometric chart, determine (a) the specific humidity, (b) the specific enthalpy and total enthalpy, (c) the wet-bulb temperature, (d) the dew-point temperature, and (e) the specific volume of the air in the room.

SOLUTION

The state of the air is specified by the given temperature and relative humidity. Other properties of the air can conveniently be found from the chart.

(a) The specific humidity is determined by drawing a horizontal line from the specified state to the right until the horizontal line intersects with the specific humidity ω axis. It is

$$\omega = 0.0105 \ \frac{\text{kg } H_2O}{\text{kg dry air}}$$

(b) The enthalpy of air is determined by drawing a line parallel to the h lines from the specific state until the line intersects the enthalpy scale. It is

$$h = 62.0 \frac{\text{kJ}}{\text{kg dry air}}$$

The total enthalpy in the room is

$$H = mh = 30 \text{ kg}\left(62.0 \ \frac{\text{kJ}}{\text{kg dry air}}\right) = \textbf{1,860 kJ}$$

(c) The wet-bulb temperature T_{wb} is determined by locating the point of the specified state between the constant T_{wb} lines on both of its sides or drawing a line parallel to the T_{wb} lines from the specified state until the line intersects the saturation line. It is read by visual interpolation to be

$$T_{wb} = \textbf{21.3°C}$$

(d) The dew-point temperature is determined by drawing a horizontal line from the specified state to the left until the horizontal line intersects the saturation line. It is read by visual interpolation to be

$$T_{dp} = \textbf{19.4°C}$$

(e) The specific volume v is determined by the same method as finding T_{wb} to locate the point of the specified state between the constant v lines on both of its sides. It is read by visual interpolation to be

$$v = \textbf{0.883} \ \frac{\textbf{m}^3}{\textbf{kg dry air}}$$

The air properties and solution steps of Example 3.2 are illustrated on the following psychrometric chart.

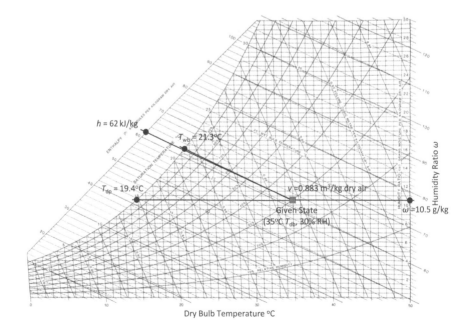

FIGURE Example 3.2

3.3 PROCESSES ON THE PSYCHROMETRIC CHART

Psychrometric process of the atmospheric air, i.e., any property change of the air at atmospheric condition, can be graphically represented by a line connecting the state points on the psychrometric chart. The line is called a process line. The basic processes aiming in one function, temperature or humidity. They are simple heating (raising the air temperature only), simple cooling (lowering the air temperature only), humidifying (increasing the moisture content of the air only), and dehumidifying (reducing the moisture content of the air only). Some processes can perform two functions, both temperature and humidity. They are: heating with humidifying (raising the air temperature and humidity simultaneously), cooling with dehumidifying (lowering the air temperature and humidity simultaneously), heating with dehumidifying (raising the air temperature but lowering air humidity simultaneously), cooling with humidifying (lowering the air temperature but raising air humidity simultaneously). Sometimes, two or more of basic processes may need to be involved for taking the air to a desired state. Figure 3.1 shows eight processes represented on the psychrometric chart.

In engineering practice, the processes in application usually are steady processes. Using mass conservative Equation (2.20) (see Chapter 2), the mass of dry air entering and exiting the process is

$$\sum_{in} \dot{m}_a = \sum_{out} \dot{m}_a$$

and the mass of moisture in the air entering and exiting the process is

$$\sum_{in} \dot{m}_w = \sum_{out} \dot{m}_w$$

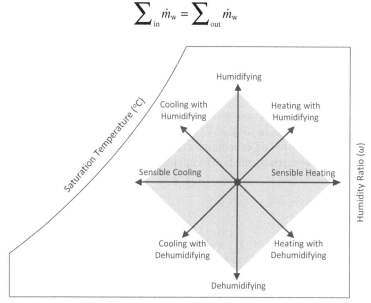

FIGURE 3.1   Processes represented on the psychrometric chart.

or

$$\sum_{in} \dot{m}_a \omega = \sum_{out} \dot{m}_a \omega$$

Using energy conservative Equation (2.24) (see Chapter 2) and neglecting kinetic and potential energies, the energy of air entering and exiting the process becomes

$$\dot{Q}_{in} + \dot{W}_{in} + \sum_{in} \dot{m}h = \dot{Q}_{out} + \dot{W}_{out} + \sum_{out} \dot{m}h$$

If there is only one air stream involved in the process, then

$$\dot{m}_{in} = \dot{m}_{out} = \dot{m}$$

$$q - w = h_{out} - h_{in}$$

Without considering any work, the heat rate of air is expressed as

$$\dot{Q} = \dot{m}(h_2 - h_1) \tag{3.11a}$$

or

$$q = h_2 - h_1 \tag{3.11b}$$

where h_1 and h_2 are the specific enthalpies of air entering and exiting the process, respectively.

3.3.1 SENSIBLE HEATING AND COOLING

When moist air is simply heated or cooled, the process is sensible heating or sensible cooling. They are simple heating and simple cooling processes. In the process, the specific humidity of the air ω remains constant. On the psychrometric chart, the processes are represented by the horizontal lines as shown in Figure 3.1.

Example 3.3

Air at 1 atm pressure, 35°C T_{db}, and 45% RH enters a cooling section of 40 cm diameter at a velocity of 10 m/s as shown. If heat is removed from the air at a rate of 700 kJ/min by the cooling water, determine (a) the mass flow rate to the section, (b) the exit enthalpy, (c) the exit temperature, (d) the exit relative humidity and specific volume, and (e) the exit velocity of the air from the section.

SOLUTION

This is the simple cooling process. The state of the air entering the cooling section is specified by the given temperature and relative humidity. Referring to Appendix A.4 Psychrometric Chart at a Pressure of 1 atm (101.325 kPa), the following properties of the air at the inlet state are found

$$h_1 = 76.14 \ \frac{kJ}{kg \ dry \ air}$$

$$v_1 = 0.8953 \ \frac{m^3}{kg \ dry \ air}$$

Since the process is a sensible cooling process, the amount of moisture in the air remains constant as the air flows through the section, i.e., $\omega_1 = \omega_2$. From the psychrometric chart, the humidity is found to be

$$\omega_1 = \omega_2 = 0.01594 \ \frac{kg \ H_2O}{kg \ dry \ air}$$

(a) The mass flow rate of the air to the section is

$$\dot{m} = \dot{m}_1 = \frac{V_1}{v_1} \frac{\pi}{4} D^2$$

$$= \frac{10 \ \frac{m}{s}}{0.8953 \ \frac{m^3}{kg \ dry \ air}} \left(\frac{\pi}{4}\right)(0.4 \ m)^2 = 1.4036 \ \frac{kg \ dry \ air}{s}$$

(b) Using Equation (3.11a), the heat removed from the air is

$$-\dot{Q}_{out} = \dot{m}(h_2 - h_1)$$

$$-\left(\frac{700}{60}\right)\frac{kJ}{s} = \left(1.4036 \ \frac{kg \ dry \ air}{s}\right)(h_2 - 76.14)\frac{kJ}{kg \ dry \ air}$$

The exit enthalpy of the air, therefore, is

$$h_2 = 67.83 \ \frac{kJ}{kg \ dry \ air}$$

(c) The exit state of the air is specified now since h_2 and ω_2 are known. Referring to Appendix A.4 Psychrometric Chart at a Pressure of 1 atm (101.325 kPa), the exit temperature of the air is found,

$$T_2 = 26.9°C$$

(d) Similarly, the exit relative humidity and the specific volume of the air are determined by using the psychrometric chart,

$$\phi_2 = \textbf{70\% RH}$$

$$v_2 = \textbf{0.8719} \; \frac{\textbf{m}^3}{\textbf{kg dry air}}$$

(e) Using Equation (2.21a) (see Chapter 2)

$$\dot{m}_1 = \dot{m}_2$$

$$\frac{\dot{V}_1}{v_1} = \frac{\dot{V}_2}{v_2}$$

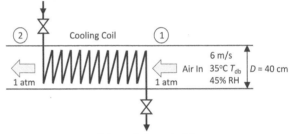

Schematic Example 3.3

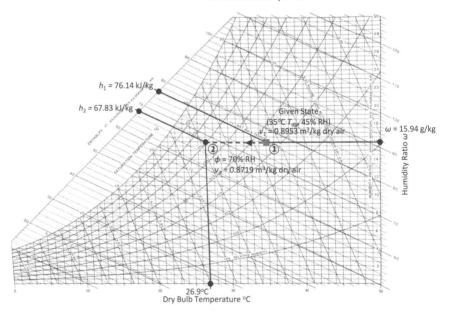

FIGURE Example 3.3

and Equation (2.13) (see Chapter 2),

$$\frac{V_1 A}{v_1} = \frac{V_2 A}{v_2}$$

The exit velocity of the air from the section is determined to be

$$V_2 = \left(\frac{v_2}{v_1}\right) V_1 = \left(\frac{0.8719 \dfrac{m^3}{kg\ dry\ air}}{0.8953 \dfrac{m^3}{kg\ dry\ air}}\right) \left(10 \frac{m}{s}\right) = 9.74 \frac{m}{s}$$

The air properties and the process line of Example 3.3 are illustrated on the following psychrometric chart.

3.3.2 HUMIDIFYING AND DEHUMIDIFYING

Humidifying air is a process to increase moisture content in air while the air dry-bulb temperature T_{db} maintains constant during the process. Dehumidifying air is just the opposite process to the humidifying of air, which is to reduce moisture content in air while the air dry-bulb temperature T_{db} remains constant. On the psychrometric chart, the processes are represented by vertical lines as shown in Figure 3.1. In application, humidifying is commonly done by spraying steam or water into the air. If spraying steam into the air, the process will move up toward the right slightly on the psychrometric chart. If spraying water into the air, the process will move up toward the left. To have the same dry-bulb temperature T_{db} at the end of the process, an additional sensible heating or cooling process is needed. When using the device called the desiccant dehumidifier for dehumidifying, the moisture content in the air will drop. However, the dry-bulb temperature T_{db} of the air may increase. To keep the same dry-bulb temperature T_{db} at the end of the process, an additional sensible cooling process is needed.

Example 3.4

Air ventilation at a volume flow rate of 22 m³/min is from the outdoors at the condition of 30°C T_{db} and 25% RH. If the room needs the air condition of 50% RH at the same temperature, determine (a) the humidification load (amount of water added) required (kg/min) and (b) the volume flow rate (m³ dry air/s) entering the room after the humidifying.

SOLUTION

This is the process of humidifying. The states of the air from the outdoors and the air in the room condition are specified by the given temperatures and relative

humidities. Assuming the water spraying does not result in the dry-bulb temperature change and referring to Appendix A.4 Psychrometric Chart at a Pressure of
1 atm (101.325 kPa), the following properties of original air and later air in humidifying are determined,

$$v_1 = 0.8675 \; \frac{m^3}{kg \; dry \; air}$$

$$\omega_1 = 0.0068 \; \frac{kg \; H_2O}{kg \; dry \; air}$$

$$h_1 = 46.6 \; \frac{kJ}{kg \; dry \; air}$$

and

$$v_2 = 0.8761 \; \frac{m^3}{kg \; dry \; air}$$

$$\omega_2 = 0.0134 \; \frac{kg \; H_2O}{kg \; dry \; air}$$

$$h_2 = 64.3 \; \frac{kJ}{kg \; dry \; air}$$

In the humidifying process, the dry air mass balance is,

$$\dot{m}_{a1} = \dot{m}_{a2} = \dot{m}_a$$

and the water mass balance is,

$$\dot{m}_{a1}\omega_1 = \dot{m}_{a2}\omega_2 + \dot{m}_w$$

$$\dot{m}_w = \dot{m}_a(\omega_1 - \omega_2)$$

The energy balance is,

$$\sum_{in} \dot{m}h = \dot{Q}_{out} + \sum_{out} \dot{m}h$$

$$\dot{Q}_{out} = \dot{m}(h_1 - h_2) - \dot{m}_w h_w$$

(a) The rate of dry air mass flow is determined to be

$$\dot{m}_a = \frac{\dot{V}_1}{v_1} = \frac{22 \; \dfrac{m^3}{min}}{0.8675 \; \dfrac{m^3}{kg \; dry \; air}} = 25.36 \; kg/min$$

The humidification load required in the process, therefore, is

$$\dot{m}_w = \dot{m}_a(\omega_2 - \omega_1)$$

$$\dot{m}_w = \left(25.36\frac{\text{kg dry air}}{\text{min}}\right)(0.0134 - 0.0068)\frac{\text{kg H}_2\text{O}}{\text{kg dry air}}$$

$$= 0.1674\frac{\text{kg H}_2\text{O}}{\text{min}}$$

(b) The volume flow rate entering the room is

$$\dot{V}_2 = v_2\dot{m}_a = \left(0.8761\frac{\text{m}^3}{\text{kg dry air}}\right)\left(25.36\frac{\text{kg dry air}}{\text{min}}\right)$$

$$= 22.22\frac{\text{m}^3}{\text{min}} = \frac{22.22\dfrac{\text{kJ}}{\text{min}}}{60\dfrac{\text{s}}{\text{min}}} = 0.3703\ \text{m}^3/\text{s}$$

The air properties and process line of Example 3.4 are illustrated in the following psychrometric chart.

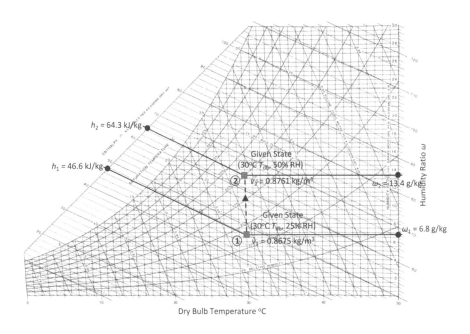

FIGURE Example 3.4

3.3.3 HEATING WITH HUMIDIFYING

Heating with the humidifying of air is a process to increase temperature and moisture in the air simultaneously. On the psychrometric chart, the process can be represented by a line extending upward to the right as shown in Figure 3.1. Humidifying normally can be accomplished by spraying steam or water into the air. In engineering practices, heating with humidifying may be accomplished by using a heating coil followed by a humidifier.

Example 3.5

A coil with a humidifier operates in a heating section at a pressure of 1 atm. The humidifier supplies the saturated steam at 100°C into the section. If air at a volume flow rate of 40 m³/min enters the section at 10°C T_{db} and 70% RH and leaves the section at 20°C T_{db} and 60% RH as shown, determine *(a)* the temperature and relative humidity of air leaving the heating coil, *(b)* the rate of heat transfer in the heating section, and *(c)* the rate of the steam added to the air in the humidifying to meet the exit air condition.

SOLUTION

This is the process of heating with humidifying. The states of air entering and leaving the heating section are completely specified by the given temperatures and relative humidities at a pressure of 1 atm. Referring to Appendix A.4 Psychrometric Chart at a Pressure of 1 atm (101.325 kPa), the following properties of the entering and exiting air are determined,

$$h_1 = 23.51 \frac{kJ}{kg \ dry \ air}$$

$$\omega_1 = \omega_2 = 0.0053 \frac{kg \ H_2O}{kg \ dry \ air}$$

$$v_1 = 0.809 \frac{m^3}{kg \ dry \ air}$$

and

$$h_3 = 42.31 \frac{kJ}{kg \ dry \ air}$$

$$\omega_3 = 0.0087 \frac{kg \ H_2O}{kg \ dry \ air}$$

The water vapor pressure corresponding to the temperature 100°C is

$$h_{g@100°C} = 2675.6 \frac{kJ}{kg \ H_2O}$$

(a) The moisture in the air remains constant $\omega_1 = \omega_2$ through the heating section, but increases from point 2 to 3 ($\omega_3 > \omega_2$) in the humidifying section. The mass flow rate of dry air is,

$$\dot{m}_a = \dot{m}_{a1} = \frac{\dot{V}_1}{v_1} = \frac{40 \dfrac{m^3}{min}}{0.809 \dfrac{m^3}{kg\ day\ air}} = 49.44 \frac{kg\ dry\ air}{min}$$

Since $Q = W = 0$, the energy balance of a steady flow on the humidifying section is

$$\dot{E}_{in} = \dot{E}_{out}$$

$$\Sigma \dot{m}_i h_i = \Sigma \dot{m}_o h_o$$

Then $\dot{m}_w h_w + \dot{m}_{a2} h_2 = \dot{m}_a h_3$

$$(w_3 - w_2)h_w + h_2 = h_3$$

h_2 is determined to be

$$h_2 = h_3 - (w_3 - w_2)h_{g@100°C}$$

$$= \left(42.31 \frac{kJ}{kg\ dry\ air}\right) - (0.0087 - 0.0053)\frac{kg\ H_2O}{kg\ dry\ air}\left(2675.6 \frac{kJ}{kg\ H_2O}\right)$$

$$= 33.21 \frac{kJ}{kg\ dry\ air}$$

Thus, the state of the air at the exit of the heating section is completely specified by $\omega_2 = 0.0053$ kg H_2O dry air and $h_2 = 33.21$ kJ/kg dry air. From the psychrometric chart, the T_{db} and RH of the state are found,

$$T_2 = \textbf{19.5°C}$$

$$\phi_2 = \textbf{38\% RH}$$

(b) The heat transfer to the air in the section, therefore, is

$$\dot{Q}_{in} = \dot{m}_a(h_2 - h_1)$$

$$= \left(49.44 \frac{kg\ dry\ air}{min}\right) \times (33.21 - 23.51)\frac{kJ}{kg\ dry\ air}$$

$$= \textbf{479.57} \frac{\textbf{kJ}}{\textbf{min}}$$

(c) Considering the conservation of mass equation in the humidifying section, the amount of water vapor added to the air in the humidifying section is determined,

$$\dot{m}_w = \dot{m}_a(\omega_3 - \omega_2)$$

$$= \left(49.44 \frac{\text{kg dry air}}{\text{min}}\right)(0.0087 - 0.0053) \frac{\text{kg H}_2\text{O}}{\text{kg dry air}}$$

$$= 0.1581 \frac{\text{kg H}_2\text{O}}{\text{min}}$$

The air properties and process lines of Example 3.5 are illustrated in the following psychrometric chart.

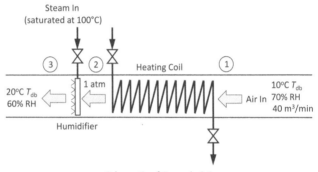

Schematic of Example 3.5

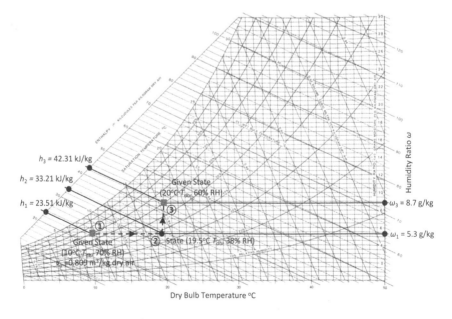

FIGURE Example 3.5

3.3.4 COOLING WITH HUMIDIFYING

Cooling by humidifying air is a process to reduce temperature and increase the moisture of the air simultaneously. On the psychrometric chart, the process is represented by a line extending upward to the left as shown in Figure 3.1. The process can be accomplished by spraying water into the air.

Example 3.6

Air at 1 atm pressure, 40°C T_{db}, and 20% RH enters an evaporative cooler and leaves with 90% RH at the same pressure. If the volume flow rate of the air is 10 m³/min and the temperature of the evaporative cooling water is close to the temperature of the exit air, determine (a) the exit temperature of the air leaving the cooler and (b) the required rate of water supply to the cooler.

SOLUTION

This is the process of cooling with humidifying. The state of air entering the cooler is specified by the given temperature and relative humidity. Referring to Appendix A.4 Psychrometric Chart at a Pressure of 1 atm (101.325 kPa), the following properties of the entering air are determined,

$$T_{wb1} = 22.04°C$$

$$\omega_1 = 0.0092 \ \frac{kg \ H_2O}{kg \ dry \ air}$$

$$v_1 = 0.9001 \ \frac{m^3}{kg \ dry \ air}$$

(a) Since the temperature of the evaporative cooling water is close to the temperature of the exit air, the evaporative cooling process follows a line of constant wet-bulb temperature. That is,

$$T_{wb2} \doteq T_{wb1} = 22.02°C$$

Knowing the wet-bulb temperature and 90% relative humidity, the dry-bulb temperature of the exit air is

$$T_{db2} = \mathbf{23.3°C}$$

$$\omega_2 = 0.0162 \ \frac{kg \ H_2O}{kg \ dry \ air}$$

(b) The mass flow rate of dry air is

$$\dot{m}_a = \dot{m}_{a1} = \frac{\dot{V}_1}{v_1} = \frac{10 \ \frac{m^3}{min}}{0.9001 \ \frac{m^3}{kg}} = 11.11 \ \frac{kg}{min}$$

The required rate of water supply to the evaporative cooler, therefore, is determined to be

$$\dot{m}_{water} = \dot{m}_{w2} - \dot{m}_{w1} = \dot{m}_a \left(\omega_2 - \omega_1 \right)$$

$$= \left(11.11 \frac{kg\ dry\ air}{min} \right) (0.0162 - 0.0092) \frac{kg\ H_2O}{kg\ dry\ air}$$

$$= 0.0778 \frac{kg\ H_2O}{min}$$

The air properties and process line of Example 3.6 are illustrated in the following psychrometric chart.

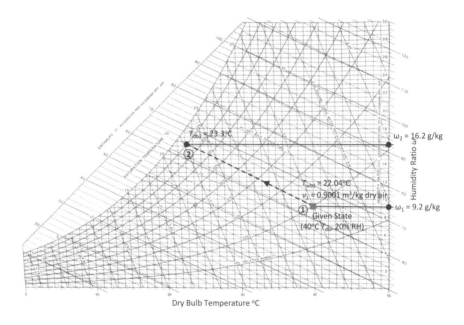

FIGURE Example 3.6

3.3.5 HEATING WITH DEHUMIDIFYING

Heating with the dehumidifying of air is a process to increase temperature and reduce the moisture of the air simultaneously during the process. A typical apparatus performing the process is a device with a desiccant. When the air comes into contact with the warm desiccant, the air is heated and moisture in the air decreases by the desiccant's absorption. On the psychrometric chart, this process is represented by a straight line extending downward to the right as shown in Figure 3.1.

Example 3.7

A rotating desiccant wheel is used to keep a room in a computer lab at a constant condition. The design requires the air leaving the wheel at 1 atm pressure with 30°C T_{db} and 20% RH. If the air entering the wheel is 20 m³/min with 25°C T_{db} and 50% RH at the same pressure, determine *(a)* the moisture rate (kg H$_2$O/min) removed from the air and *(b)* the rate of heat (kW) added to the air in the process.

SOLUTION

This is the process of heating with dehumidifying. The states of the air entering and exiting the rotating desiccant wheel are specified by the given temperatures and relative humidities. Referring to Appendix A.4 Psychrometric Chart at a Pressure of 1 atm (101.325 kPa), the following properties of the air entering air and leaving the wheel are determined,

$$v_1 = 0.8581 \frac{m^3}{kg \ dry \ air}$$

$$h_1 = 50.8 \frac{kJ}{kg \ dry \ air}$$

$$\omega_1 = 0.01 \frac{kg \ H_2O}{kg \ dry \ air}$$

and

$$v_2 = 0.8661 \frac{m^3}{kg \ dry \ air}$$

$$h_2 = 43.5 \frac{kJ}{kg \ dry \ air}$$

$$\omega_2 = 0.0053 \frac{kg \ H_2O}{kg \ dry \ air}$$

The water mass balance in the process is:

$$\dot{m}_{a1}\omega_1 = \dot{m}_{a2}\omega_2 + \dot{m}_w$$

$$\dot{m}_w = \dot{m}_a(\omega_1 - \omega_2)$$

The mass flow of the dry air is

$$\dot{m}_a = \dot{m}_{a1} = \frac{\dot{V}_1}{v_1} = \frac{20 \ \frac{m^3}{min}}{0.8581 \ \frac{m^3}{kg \ dry \ air}} = 23.32 \ \frac{kg \ dry \ air}{min}$$

(a) The moisture removed from the air is determined to be

$$\dot{m}_w = \dot{m}_{w2} - \dot{m}_{w1} = \dot{m}_a (\omega_2 - \omega_1)$$

$$= \left(23.32 \, \frac{\text{kg dry air}}{\text{min}} \right)(0.01 - 0.0053)\frac{\text{kg H}_2\text{O}}{\text{kg dry air}}$$

$$= 0.1096 \, \frac{\text{kg}}{\text{min}}$$

(b) Using Equation (3.11a), the rate of heat added in the air in the process is

$$\dot{Q} = \dot{m}(h_2 - h_1) = \left(23.32 \, \frac{\text{kg dry air}}{\text{min}} \right)(50.8 - 43.5)\frac{\text{kJ}}{\text{kg dry air}}$$

$$= 170.24 \, \frac{\text{kJ}}{\text{min}} = \mathbf{2.8 \ kW}$$

The air properties and process line of Example 3.7 are illustrated in the following psychrometric chart.

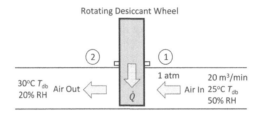

Schematic of Example 3.7

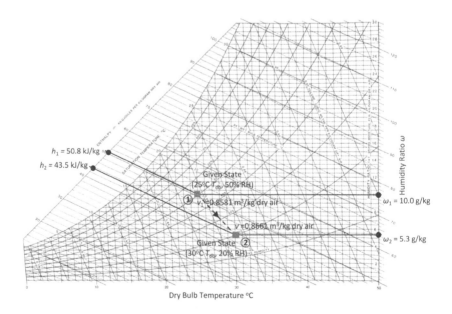

FIGURE Example 3.7

3.3.6 COOLING WITH DEHUMIDIFYING

Cooling with the dehumidifying of air is a process to reduce the temperature and moisture of the air simultaneously during the process, which is a common process in air conditioning. Typically, the process is accomplished by using the cooling coil with chilled water or refrigerant flowing inside the coil. On the psychrometric chart, the process is represented by a line extending downward to the left as shown in Figure 3.1.

Example 3.8

Air at a volume flow rate of 15 m³/min enters an air conditioner at 1 atm pressure, 30°C T_{db}, and 60% RH as shown. If the air leaves as saturated air at 15°C and the same pressure, determine (a) the moisture rate (kg H₂O/min), (b) the rate of total heat (kW), and (c) the rate of latent heat and sensible heat (kW) removed from the air in the process.

SOLUTION

This is the process of cooling with dehumidifying. The state of the air entering the air conditioner is specified by the given temperature and relative humidity and the state of the leaving air is specified by the saturated temperature. Referring to Appendix A.4 Psychrometric Chart at a Pressure of 1 atm (101.325 kPa), the following properties of the air entering and leaving the air conditioner are determined,

$$v_1 = 0.8805 \ \frac{m^3}{kg \ dry \ air}$$

$$h_1 = 71.2 \ \frac{kJ}{kg \ dry \ air}$$

$$\omega_1 = 0.0161 \ \frac{kg \ H_2O}{kg \ dry \ air}$$

and

$$h_2 = 42.0 \ \frac{kJ}{kg \ dry \ air}$$

$$\omega_2 = 0.0107 \ \frac{kg \ H_2O}{kg \ dry \ air}$$

In the cooling with dehumidifying process, the dry air mass balance is,

$$\dot{m}_{a1} = \dot{m}_{a2} = \dot{m}_a$$

and the water mass balance is,

$$\dot{m}_{a1}\omega_1 = \dot{m}_{a2}\omega_2 + \dot{m}_w$$

$$\dot{m}_w = \dot{m}_a(\omega_1 - \omega_2)$$

The total heat removed from the air in the process becomes

$$\dot{Q}_{out} = \dot{m}(h_1 - h_2)$$

and

$$\dot{Q}_{out} = \dot{Q}_s + \dot{Q}_l$$

(a) The mass flow rate entering the air conditioner is

$$\dot{m}_a = \frac{\dot{V}_1}{v_1} = \frac{15 \, \frac{m^3}{min}}{0.8805 \, \frac{m^3}{kg \; dry \; air}} = 17.04 \, \frac{kg \; dry \; air}{min}$$

The moisture rate removed from the air is determined to be

$$\dot{m}_w = \left(17.04 \, \frac{kg \; dry \; air}{min}\right)(0.0161 - 0.0107)\frac{kg \; H_2O}{kg \; dry \; air}$$

$$= 0.092 \, \frac{kg \; H_2O}{min}$$

(b) Using Equation (3.11a), the total heat removed from the air in the process is obtained,

$$\dot{Q}_{out} = \dot{m}(h_1 - h_2)$$

$$= \left(17.04 \, \frac{kg \; dry \; air}{min}\right)(71.2 - 42.0)\frac{kJ}{kg \; dry \; air}$$

$$= 497.57 \, \frac{kJ}{min} = \frac{497.57 \, \frac{kJ}{min}}{60 \, \frac{s}{min}} = 8.2928 \; kW$$

(c) Using the enthalpies shown in the psychrometric chart, the rate of latent heat and sensible heat is determined,

$$\dot{Q}_l = \dot{m}h_l = \left(17.04 \, \frac{kg \; dry \; air}{min}\right)(71.2 - 57.6)\frac{kJ}{kg \; dry \; air}$$

$$= 231.74 \, \frac{kJ}{min} = \frac{231.74 \, \frac{kJ}{min}}{60 \, \frac{s}{min}} = 3.8624 \; kW$$

$$\dot{Q}_s = \dot{m}h_s = \left(17.04 \; \frac{\text{kg dry air}}{\text{min}}\right)(57.6 - 42.0) \; \frac{\text{kJ}}{\text{kg dry air}}$$

$$= 265.82 \; \frac{\text{kJ}}{\text{min}} = \frac{265.82 \; \dfrac{\text{kJ}}{\text{min}}}{60 \; \dfrac{\text{s}}{\text{min}}} = \textbf{4.4304 kW}$$

The air properties and process line of Example 3.8 are illustrated in the following psychrometric chart.

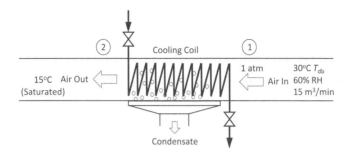

Schematic of Example 3.8

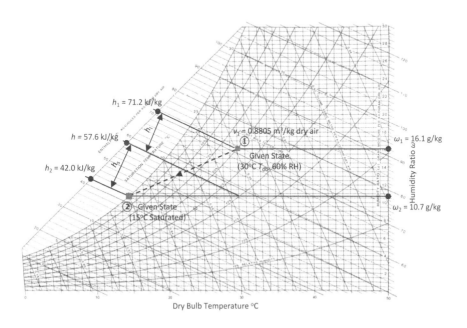

Dry Bulb Temperature °C

FIGURE Example 3.8

3.3.7 ADIABATIC MIXING OF AIRSTREAMS

Adiabatic mixing of airstreams is a popular practice in engineering applications. In a large air conditioning system, the return air from the air-conditioned area and the fresh air from the outdoors may be mixed to form an air supply. The mixed air is obtained by simply merging the two airstreams as shown in Figure 3.2.

The mixing process normally does not involve work interaction. Changes in kinetic and potential energies between the airstreams, if any, are negligible. The heat transfer of the airstreams with the surroundings usually is small during the mixing process. Therefore, the mixing process can be assumed to be adiabatic. The mass and energy balances for the adiabatic mixing of two airstreams are expressed as

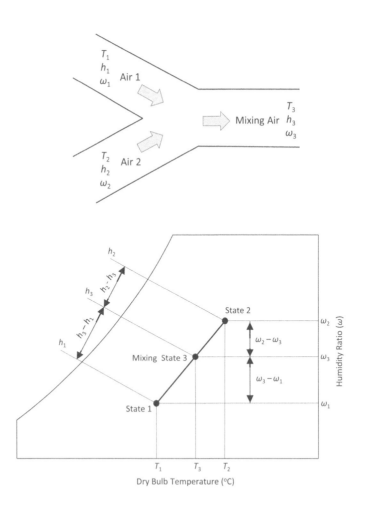

FIGURE 3.2 Mixed air by two airstreams.

$$\text{Dry air mass balance: } \dot{m}_{a1} = \dot{m}_{a2} = \dot{m}_{a3} \tag{3.12a}$$

$$\text{Water vapor mass balance: } \dot{m}_{a1}\omega_1 = \dot{m}_{a2}\omega_2 = \dot{m}_{a3}\omega_3 \tag{3.12b}$$

$$\text{Energy balance: } \dot{m}_{a1}h_1 + \dot{m}_{a2}h_2 = \dot{m}_{a3}h_3 \tag{3.12c}$$

Replacing $\dot{m}_{a3}$ in the above relations, the following equation is obtained,

$$\frac{\dot{m}_{a1}}{\dot{m}_{a2}} = \frac{\omega_2 - \omega_3}{\omega_3 - \omega_1} = \frac{h_2 - h_3}{h_3 - h_1} \tag{3.13}$$

This equation has a geometric interpretation on the psychrometric chart shown in Figure 3.2. After two airstreams at two different states (states 1 and 2) are mixed adiabatically, the state of the mixture (state 3) lies on the straight line connecting states 1 and 2. The ratio of the distances 2–3 and 3–1 is equal to the ratio of mass flow rates $\dot{m}_{a1}$ and $\dot{m}_{a2}$. Using Equation (2.31b) (see Chapter 2), Equation (3.12c) becomes,

$$\dot{m}_{a1}c_{p1}T_1 + \dot{m}_{a2}c_{p2}T_2 = \dot{m}_{a3}c_{p3}T_3 \tag{3.14}$$

In engineering applications, the specific heat of an airstream is assumed to have no big change, i.e., $c_{p1} = c_{p2} = c_{p3}$. Therefore, the temperature of the mixed air is determined to be

$$T_3 = \frac{\dot{m}_{a1}T_1 + \dot{m}_{a2}T_2}{\dot{m}_{a3}} \tag{3.15}$$

Example 3.9

Two airstreams are mixed steadily and adiabatically. The first airstream at a rate of 20 m³/min enters at 30°C T_{db} and 25% RH. The second airstream at a rate of 25 m³/min enters at 35°C T_{db} and 60% RH. If the mixing process is at a constant pressure of 1 atm, determine (a) the enthalpy and the specific humidity, (b) the dry-bulb temperature, (c) the relative humidity and the specific volume, and (d) the volume flow rate (m³/min) of the mixed air.

SOLUTION

The states of the two entering air streams are specified by the given temperatures and relative humidities. Referring to Appendix A.4 Psychrometric Chart at a Pressure of 1 atm (101.325 kPa), the following properties of the two entering airstreams are determined,

$$v_1 = 0.8675 \; \frac{m^3}{kg \; dry \; air}$$

$$\omega_1 = 0.0068 \; \frac{kg \; H_2O}{kg \; dry \; air}$$

$$h_1 = 46.6 \frac{kJ}{kg\ dry\ air}$$

and

$$v_2 = 0.9031 \frac{m^3}{kg\ dry\ air}$$

$$\omega_2 = 0.0215 \frac{kg\ H_2O}{kg\ dry\ air}$$

$$h_2 = 90.4 \frac{kJ}{kg\ dry\ air}$$

The mass flow rates of the dry air of the two airstreams are, respectively,

$$\dot{m}_{a1} = \frac{\dot{V}_1}{v_1} = \frac{20 \frac{m^3}{min}}{0.8675 \frac{m^3}{kg\ dry\ air}} = 23.06 \frac{kg\ dry\ air}{min}$$

and

$$\dot{m}_{a2} = \frac{\dot{V}_2}{v_2} = \frac{25 \frac{m^3}{min}}{0.9031 \frac{m^3}{kg\ dry\ air}} = 27.68 \frac{kg\ dry\ air}{min}$$

Mass of the mixed air is a summation of the two entering airstreams,

$$\dot{m}_{a3} = \dot{m}_{a1} + \dot{m}_{a2} = (23.06 + 27.68) \frac{kg\ dry\ air}{min}$$

$$= 50.74 \frac{kg\ dry\ air}{min}$$

(a) Using Equation (3.13),

$$\frac{\dot{m}_{a1}}{\dot{m}_{a2}} = \frac{\omega_2 - \omega_3}{\omega_3 - \omega_1} = \frac{h_2 - h_3}{h_3 - h_1}$$

$$\frac{23.06 \frac{kg\ dry\ air}{min}}{27.68 \frac{kg\ dry\ air}{min}} = \frac{(0.0215 - \omega_3)\frac{kg\ H_2O}{kg\ dry\ air}}{(\omega_3 - 0.0068)\frac{kg\ H_2O}{kg\ dry\ air}} = \frac{(90.4 - h_3)\frac{kJ}{kg\ dry\ air}}{(h_3 - 46.6)\frac{kJ}{kg\ dry\ air}}$$

and manipulating the above equation, specific humidity and enthalpy of the mixed air are obtained as

$$h_3 = 70.49 \ \frac{kJ}{kg \ dry \ air}$$

$$\omega_3 = 0.0148 \ \frac{kg \ H_2O}{kg \ dry \ air}$$

The state of the mixed air is specified by these two properties. Other properties of the mixed air can be calculated by equations or determined from the psychrometric chart.

(b) Using Equation (3.15), the dry-bulb temperature is determined to be

$$T_3 = \frac{\dot{m}_{a1}T_1 + \dot{m}_{a2}T_2}{\dot{m}_{a3}}$$

$$= \frac{\left(23.06 \ \frac{kg \ dry \ air}{min}\right)30°C + \left(27.68 \ \frac{kg \ dry \ air}{min}\right)35°C}{50.74 \ \frac{kg \ dry \ air}{min}}$$

$$= 32.73°C$$

(c) The relative humidity and specific volume are determined from the psychrometric chart by the state point of the mixed air

$$\phi_3 = 47\%$$

$$v_3 = 0.8865 \ \frac{m^3}{kg \ dry \ air}$$

(d) The volume flow rate of the mixture stream becomes

$$\dot{V}_3 = \dot{m}_{a3}v_3 = \left(50.74 \ \frac{kg \ dry \ air}{min}\right)\left(0.8865 \ \frac{m^3}{kg \ dry \ air}\right)$$

$$= 44.98 \ m^3/min$$

The air properties and process lines of Example 3.9 are illustrated in the following psychrometric chart.

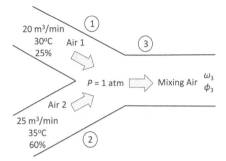

Schematic of Example 3.9

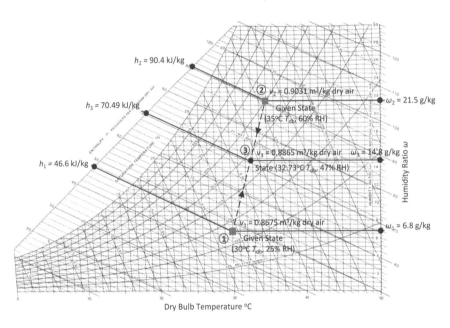

FIGURE Example 3.9

4 Heating and Cooling Coils

4.1 INTRODUCTION

A coil is a device of thermal energy exchange. The coil mainly composed of tubes with a small diameter and thin wall thickness. The tubes are bundled up with other components to work as a heating coil or cooling coil. The coil transfers heat between fluids unmixed. One fluid flows inside the tube and another fluid flows outside the tube. In application of heating and cooling of air passing through coils, air flows outside the tubes and the fluid, such as steam, water, or refrigerant, flows inside the tubes. A typical coil is shown in Figure 4.1.

When a fluid at a higher temperature flows inside the tubes, the coil is called a heating coil. The heat transfer between two fluids is a heating process. When the fluid at a lower temperature flows inside the tubes, the coil is called a cooling coil. The heat transfer between two fluids is a cooling process. Finned tubes are commonly used in heating and cooling coils to enhance the heat transfer effect. The coil tubes are connected to the manifolds, which are called the distribution header and discharge header for fluid entering and exiting the coil and circulating into the fluid piping system as well.

4.2 COIL TYPES AND CONSTRUCTION

4.2.1 COIL TYPES

Types of heating and cooling coils can be classified in various ways, such as the type of fluid flowing inside the coil tubes, coil tube materials, and coil tubes with or without fins.

Water Coil

A water coil is where water flows inside the coil tubes. The water coil can be a heating coil or a cooling coil depending on the water temperature. A water coil is the heating coil if hot water flows inside the coil tubes. Otherwise, a water coil is a cooling coil, also called a chilled water coil, if chilled water flows inside the coil tubes.

Refrigerant Coil

A refrigerant coil is the cooling coil. The low-temperature refrigerant flows inside the coil tubes. Refrigerants commonly used in air conditioning coils are R134a and R404a.

Steam Coil

A steam coil is the heating coil. The steam flows inside the coil tubes. The application of the steam coil is more complicated than the hot water coil since the method

DOI: 10.1201/9781003289326-4

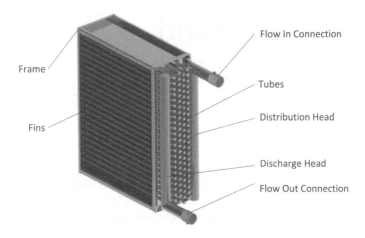

FIGURE 4.1   A typical coil. (Courtesy of Marlo Heat Transfer Solutions, 2022.)

involves draining the steam condensate out of the coil to prevent the condensate from clogging the coil tubes.

Heat Recovery Coil

A heat recovery coil is a special coil. The heat is from waste heat sources, such as from the jacket water of diesel engines or the exhaust gas of gas turbines. The jacket water and exhaust gas flow inside the coil tubes.

Coils can be constructed from different materials, such as copper or stainless steel, and with or without fins, which depends on the application conditions and economic considerations. Table 4.1 shows different materials in coil construction.

4.2.2 FINNED TUBE COILS

Coils are used to accomplish heat transfer between fluids inside and outside the tubes. Finned tube coils are widely used in engineering practices since the fins on the coil tubes greatly enhance the heat transfer rate between fluids as described in Section 2.3.2 in Chapter 2. In general, fins are estimated to have 65–70% of the total heat transfer rate on a finned tube coil.

Fin Materials

Fins can be made from a variety of materials. The selection of fin materials depends on the coil performance, working environment, corrosion condition, and economic consideration, such as

- Desired heat transfer performance.
- Compatibility of the material with the fluid.
- Frequency and aggressiveness of coil cleaning.

TABLE 4.1
Coil Materials

	Tubes	Fins	Casing	Connections
Materials	Copper: Offers the best heat transfer and is the most commonly used.	Aluminum: Offers the lowest cost and is the most commonly used.	Aluminum: Offers the lowest cost and is the most commonly used.	Copper, brass, stainless steel, and steel
	Stainless Steel: Offers high heat transfer rate and is used for high-pressure application.	Copper: Commonly used in corrosive environments.	Copper: Commonly used in corrosive environments.	
	Steel: Commonly used for high-pressure steam application.	Stainless steel: Used in corrosive or food-grade environments.	Stainless: Used in corrosive or food-grade environments.	
Sizes (mm)	DN10, DN15, and DN16 are commonly used. DN16 mm tube is the most popular used because it offers many wall thickness options.			From DN15 to DN100

Aluminum, copper, and stainless steel are the most popular materials as shown in Table 4.1.

Fin Types

In general, there are two kinds of fin types. One is called the plate fin and another is called the spiral fin or the helix-shaped fin. There is a significant difference in construction between the two types. The plate fin involves multiple tubes passing through a common fin plate, while the spiral fin involves each tube being wrapped by fins independently.

Plate Fins

In a plate finned tube coil, tubes are inserted through a series of fin plates as shown in Figure 4.2.

The plate fins can be made using a variety of patterns, such as flat or corrugated patterns. The choice of pattern is due to the consideration of flow turbulent enhancement and coil cleaning convenience.

Spiral Fins

In a spiral finned tube coil, the spiral fins are wrapped around an individual tube as shown in Figure 4.3(a). The spiral fins have good fin-to-tube contact. Some methods of the fin-to-tube attachment are shown in Figure 4.3(b).

Fin Density Variability

Coils with plate fins allow for a wide array of fin density, which has a range from 1 to 25 fins per 25.4 mm. Coils with the spiral fins tend to be less dense, which has a

FIGURE 4.2 Plate finned tubes. (Adapted from Super Radiator Coil, 2022.)

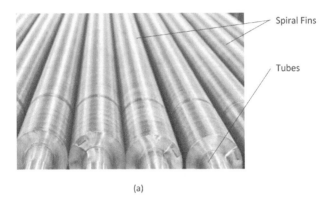

(a)

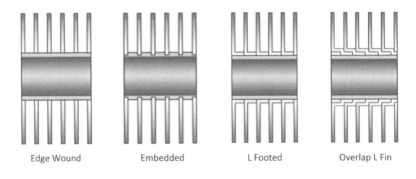

| Edge Wound | Embedded | L Footed | Overlap L Fin |

(b)

FIGURE 4.3 The spiral finned tubes. (a) Spiral finned tubes. (b) Fin to tube attachment. (Adapted from Super Radiator Coil, 2022.)

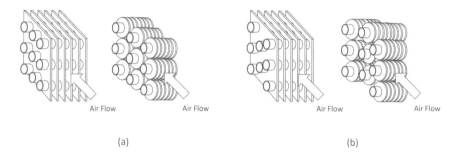

Air Flow   Air Flow   Air Flow   Air Flow

(a)   (b)

FIGURE 4.4   Tube arrangement of coils. (a) In-line. (b) Staggered.

range from 4 to 13 fins per 25.4 mm. Some spiral fins with very low fin height may achieve a larger fin density.

Heat Transfer Performance

The plate fin coil provides a better heat transfer rate on the air side than the spiral fin because the plate finned coil provides a bigger heat-transfer surface area compared to the spiral finned coil in the same flow passage.

4.2.3   TUBE ARRANGEMENT OF COILS

Tubes in a coil are typically arranged in two types: in-line or staggered along the direction of the air approaching the coil as shown in Figure 4.4.

Figure 4.5 shows some fin-to-tube parameters of the in-line coils and the staggered coils with plate finned tubes and spiral finned tubes, respectively. The parameters are necessary in the design and selection of the finned tube coils.

4.3   FLOW TYPES

4.3.1   INTERNAL FLOW AND EXTERNAL FLOW

In the heating and cooling of air passing through coils, air is an external flow and fluid inside the coils is an internal flow. Figure 4.6 shows schematics of the external flow passing a tube bank of in-line and staggered arrangement and the internal flow in laminar and turbulent types.

4.3.2   LAMINAR FLOW AND TURBULENT FLOW

The fluid flowing inside and outside of coils can be either a laminar flow or turbulent flow. The flow type depends on the tube size, fluid type, and flow velocity. The flow being laminar or turbulent is identified by the Reynolds number, which is described in Section 2.2.3 (see Chapter 2),

$$\mathrm{Re} = \frac{v_{\mathrm{avg}}D}{v} = \frac{\rho v_{\mathrm{avg}}D}{\mu}$$

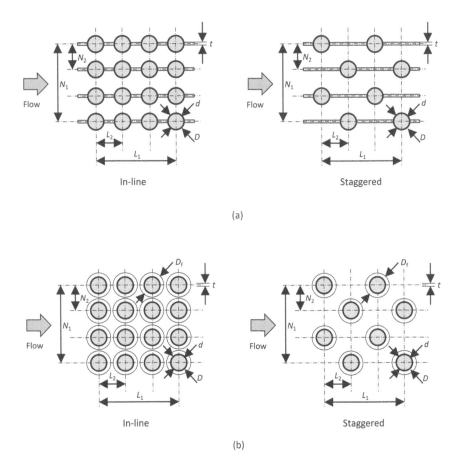

FIGURE 4.5 Fin-to-tube parameters of coils. (a) Plate finned tubes. (b) Spiral finned tubes.

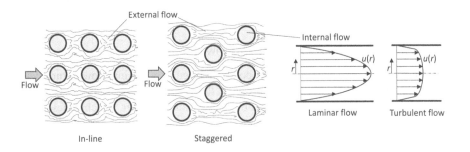

FIGURE 4.6 Schematics of external, internal, laminar, and turbulent flows.

For an internal flow, the criteria are

Re ≤ 2,300 laminar flow
2,300 ≤ Re ≤ 4,000 transitional flow
Re ≥ 4,000 turbulent flow

For an external flow passing over the coil tube, the criteria are:

Re ≤ 2 × 10⁻⁵ laminar flow
2 × 10⁻⁵ ≤ Re ≤ 2 × 10⁻⁶ transitional flow
Re ≥ 2 × 10⁻⁶ turbulent flow

Typically, the laminar flow has less fluid friction loss. Consequently, the flow consumes less power than turbulent flow. However, turbulent flow has a higher heat transfer coefficient than laminar flow. An apparatus with turbulent flow may have a smaller size than that with the laminar flow. In engineering practices, flow is commonly expected to be turbulent.

Example 4.1

Saturated water at a temperature of 25°C ($\mu = 0.891 \times 10^{-3}$ kg/m.s) enters a coil with thin copper tubes in a diameter of 2.5 cm at a volume flow rate of 0.11 m³/min. The length of the tube is 2 m. If there are 100 of the same tubes in the coil and the pump efficiency is 92%, determine (a) the flow type of the water in the coil tubes, (b) the pump power (kW) required to overcome the pressure loss for maintaining the flow in the coil.

SOLUTION

Referring to Appendix A.5 Properties of Steam and Compressed Water, the water density at the given temperature is found,

$$\rho = 997.0 \ \frac{kg}{m^3}$$

(a) The fluid velocity and the Reynolds number in the pipe is determined to be

$$V = \frac{\dot{V}}{A} = \frac{\dot{V}}{\pi\left(\dfrac{D^2}{4}\right)} = \frac{0.11\left(\dfrac{m^3}{min}\right)}{\pi\left[\dfrac{(0.025 \ m)^2}{4}\right]} = 3.7348 \ \frac{m}{s}$$

$$Re = \frac{\rho V D}{\mu} = \frac{\left(997.0 \ \dfrac{kg}{m^3}\right)\left(3.7348 \ \dfrac{m}{s}\right)(0.025 \ m)}{0.891 \times 10^{-3} \ \dfrac{kg}{m.s}}$$

$$= \mathbf{1.0448 \times 10^5 > 4{,}000}$$

Since the Reynolds number is bigger than 4,000, the fluid flow in the coil tubes is in turbulent.

(b) Referring to Appendix A.6 The Moody Chart, the roughness of copper material is 0.0015 mm. Therefore, the relative roughness is

$$\frac{\varepsilon}{D} = \frac{0.0015 \text{ mm}}{25 \text{ mm}} = 0.00006$$

Using Equation (2.39) (see Chapter 2),

$$\frac{1}{\sqrt{f}} = -1.8 \log \left[\frac{6.9}{\text{Re}} + \left(\frac{\varepsilon/D}{3.7} \right)^{1.11} \right]$$

or from the Moody chart, the friction factor is found,

$$f = 0.0179$$

Using Equation (2.37) (see Chapter 2), the pressure drop through the tube is

$$\Delta P = f \frac{L}{D} \frac{\rho V^2}{2} = 0.0179 \left(\frac{2 \text{ m}}{0.025 \text{ m}} \right) \left[\frac{\left(997.0 \frac{\text{kg}}{\text{m}^3} \right) \left(3.7348 \frac{\text{m}}{\text{s}} \right)^2}{2} \right]$$

$$= 9,957.3 \frac{\text{kg}}{\text{m.s}^2} = \left(9,957.3 \frac{\text{kg}}{\text{m.s}^2} \right) \left(\frac{1 \frac{\text{kN}}{\text{m}^2}}{1,000 \frac{\text{kg.m}}{\text{s}^2}} \right) \left(\frac{1 \text{ kPa}}{1 \frac{\text{kN}}{\text{m}^2}} \right)$$

$$= 9.9573 \text{ kPa}$$

Using Equation (2.48b) (see Chapter 2), the pump power required to overcome the pressure loss, therefore, is determined to be

$$\dot{W} = \frac{\dot{V} \Delta P}{\eta_{\text{pump}}} = \frac{100 \left(0.11 \frac{\text{m}^3}{\text{min}} \right)}{60 \frac{\text{s}}{\text{min}}} (9.9573 \text{ kPa})$$

$$= 1.98 \frac{\text{kPa.m}^3}{\text{s}} \left(\frac{1 \text{ kW}}{\frac{\text{kPa.m}^3}{\text{s}}} \right) = \textbf{1.98 kW}$$

4.3.3 FORCED INCOMPRESSIBLE FLOW

Liquids, such as chilled water and refrigerant are incompressible fluids. Steam and air are theoretically considered compressible fluids. In engineering practice, however, the flow of steam and air is usually treated as incompressible flow. Whether the flow of steam or air can be treated as incompressible flow depends on the value of the Mach number, Ma, which is defined as:

$$\text{Ma} = \frac{V}{c} = \frac{\text{Speed of flow}}{\text{Speed of sound}} \tag{4.1}$$

where c is the local speed of sound whose value is 346 m/s of air at room temperature at sea level. Ma = 1.0 is a sonic flow. When Ma < 0.3, the compressibility effects on the steam and air can be neglected. In engineering practice, the flow velocity of the steam moving in the coil tube and the air passing through the coil is typically lower than 0.3 of the Ma. Therefore, the flow of the steam and air in the heating and cooling of air passing through coils are considered forced incompressible flows.

4.3.4 FLOW DIRECTIONS OF FLUIDS

Flow directions of fluids inside and outside the tube will have a significant effect on the heat transfer between the fluids. There are three flow types based on the flow directions: parallel flow, counter flow, and cross flow.

4.3.4.1 Parallel Flow

Parallel flow, also called concurrent flow, means that the fluids inside and outside the tube have the same flow directions. The flow is characterized by the temperature difference between two fluids being large at the inlet end and becoming small at the outlet end as shown in Figure 4.7. This is a fairly inefficient flow pattern in heat transfer since the highest outlet temperature of cold flow is always lower than the exit temperature of hot flow. Some energy from the hot fluid is wasted.

4.3.4.2 Counter Flow

The counter flow means that the fluids inside and outside the tube have the opposite flow directions. The flow is characterized by the temperature difference between two fluids being fairly even during the process. The outlet temperature of the cold fluid can be greater than the outlet temperature of the hot fluid and approach the inlet temperature of the hot fluid as shown in Figure 4.8. The energy of the hot fluid in the counter flow is effectively used. The counter flow, therefore, has a higher efficiency of heat transfer than the parallel flow in the same flow conditions.

4.3.4.3 Cross Flow

In engineering applications, the counter flow may be hard to reach, particularly for finned tube coils. Cross flow is a practical alternative. In cross flow, the flow directions of the fluids inside and outside the tube are perpendicular as shown in Figure 4.9. Cross flow is a common type in the heating and cooling of air passing through coils.

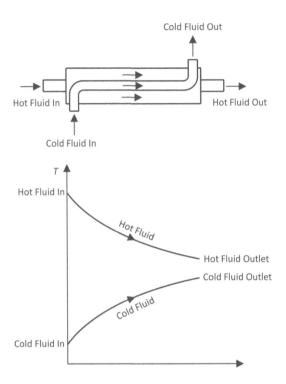

FIGURE 4.7   Parallel flow.

In the heat transfer calculation of the cross-flow coils, equivalent temperature difference ΔT_{eq} is used. ΔT_{eq} is the temperature difference modified from the log mean temperature difference (LMTD) of the counter flow, which is expressed as

$$\Delta T_{eq} = F\Delta T_{lm,cf} \qquad (4.2)$$

where F is the correction factor depending on the flow configuration and the temperatures of fluids entering and exiting the coil. $\Delta T_{lm,cf}$ is the LMTD of the counter flow at the same coil configuration and determined by Equation (2.70) described in Section 2.3.3 (see Chapter 2). The value of F is less than unity, which is a measure of the deviation for the cross flow from the counter flow.

Example 4.2

A staggered cross flow coil has 50 plate finned tubes. The tube internal diameter is 0.95 cm and the length is 60 cm. Hot water flowing inside the tubes at a rate of 0.8 kg/s has an entering temperature of 90°C and a exiting temperature of 68°C. The specific heat of the water at the average temperature is 4,1962 J/kg.K. Air passing through outside the tubes is heated from 20°C to 30°C. The correction factor

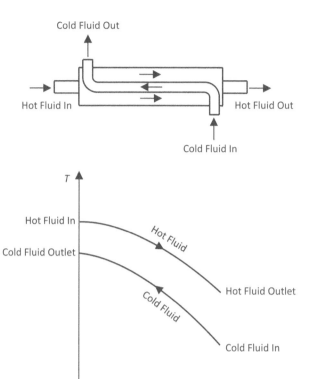

FIGURE 4.8 Counter flow.

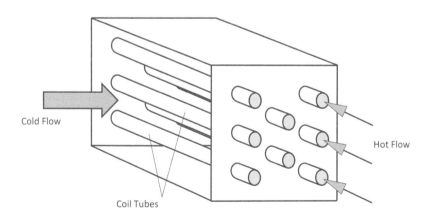

FIGURE 4.9 Cross flow arrangement.

of the counter flow is 0.99. Based on the inner surface area of the tubes, determine *(a)* the rate of heat transfer (kW) and *(b)* the overall heat transfer coefficient U_i from the inner side of the tube in steady flow.

SOLUTION

(a) Using Equation (2.68) (see Chapter 2), the rate of heat transfer in the coil tubes is determined to be

$$\dot{Q} = \dot{m}c_p \left(T_{in} - T_{out} \right)$$

$$= \left(0.8 \, \frac{kg}{s} \right) \left(4.1962 \, \frac{kJ}{kg.°C} \right) (90 - 68)°C$$

$$= 73.85 \, \frac{kJ}{s} = \mathbf{73.85 \ kW}$$

(b) The heat transfer area is the inner surface of total tubes $n = 50$, i.e.,

$$A_i = n\pi DL = 50\pi \left(0.0095 \text{ m} \right)\left(0.6 \text{ m} \right) = 0.8953 \text{ m}^2$$

Using Equation (2.62) (see Chapter 2) and Equation (4.2), the rate of heat transfer becomes

$$\dot{Q} = U_i A_i F \Delta T_{lm,cf}$$

the overall heat transfer coefficient U of the coil is determined by

$$U_i = \frac{\dot{Q}}{A_i F \Delta T_{lm,cf}}$$

The temperature differences of the counter-flow arrangement are

$$\Delta T_1 = T_{h,in} - T_{c,out} = (90 - 30)°C = 60°C$$

$$\Delta T_2 = T_{h,out} - T_{c,in} = (68 - 20)°C = 48°C$$

The LMTD of the counter flow is

$$\Delta T_{lm,cf} = \frac{\Delta T_1 - \Delta T_2}{\ln(\Delta T_1)/(\Delta 2)} = \frac{(60 - 48)°C}{\ln(60°C)/(48°C)} = 53.78°C$$

Finally, the overall heat transfer coefficient U_i is

$$U_i = \frac{73.85 \text{ kW}}{\left(0.8953 \text{ m}^2 \right)0.99 \left(53.78°C \right)} = \mathbf{1.5493} \, \frac{\mathbf{kW}}{\mathbf{m^2.K}}$$

4.4 COIL APPLICATIONS

4.4.1 Coils in Condenser and Evaporator

Coils are widely used in air conditioning (AC or A/C). They can be found in buildings, aircraft, ships, vehicles, etc. Figure 4.10 shows coils in a typical air conditioner.

In an air conditioner or an air conditioning system, two coils are installed within the indoor evaporator and outdoor condenser, respectively. The refrigerant flowing inside the coils changes phase from liquid to vapor in the evaporator by taking the heat of indoor air and from vapor to liquid in the condenser by releasing the heat outdoors. The air passing through the evaporator coil is a cooling process. The coils in the evaporator and condenser are basically the same. They are named from their functions.

4.4.2 Split-System Air Conditioning

Split-system air conditioning mainly consists of two main units installed separately. One called the indoor unit is located inside the building and another called the outdoor unit is located outside the building. The evaporator is held in the indoor unit. The condenser and compressor are held in the outdoor unit. The coils in the condenser and the evaporator are connected through copper tubes. The advantage of split-system air conditioning is the flexibility of the unit locations. Split-system air conditioning is suitable for small spaces, such as commercial offices, apartments, and residential homes. Figure 4.11 shows a schematic of the typical split-system air conditioning.

Split-system air conditioning can use a heat pump to work for both heating and cooling. In hot weather, the heat pump works in cooling mode to cool the room air. In cold weather, the heat pump works in heating mode to warm the room air.

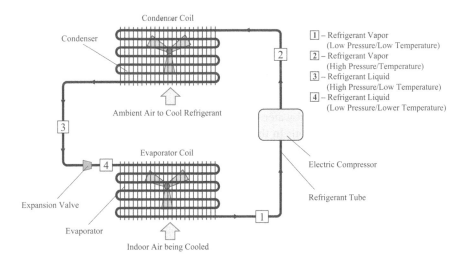

FIGURE 4.10 Coils in a typical air conditioner.

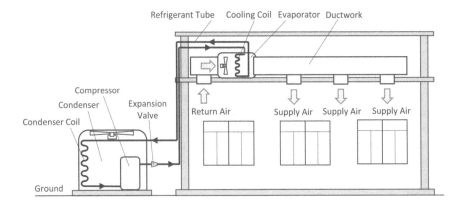

FIGURE 4.11 A schematic of the typical split-system air conditioning.

4.4.3 CENTRAL AIR CONDITIONING

A central air conditioning system, as it is named, has a hub. The hub is the air handling unit (AHU) that houses the heating and cooling coils. In the heating process, steam or hot water generated from a boiler flows inside the heating coil. In the cooling process, chilled water generated from a chiller flows inside the cooling coil. From the AHU, the conditioned air, cold in summer and warm in winter, is supplied to the air-conditioned spaces through ductwork. Central air conditioning is suitable to be used for large areas, such as tall buildings, industrial parks, and university campuses. The main advantage of central air conditioning is that numerous standalone air conditioning systems and air conditioners required for individual areas can be eliminated. Operation of central air conditioning has higher energy efficiency, lower operation and maintenance costs, and problems are quick to troubleshoot.

4.4.3.1 Chilled Water Style

In the central air conditioning system of the chilled water style, there are two cold fluid loops. One is the chilled water loop and another is the refrigerant loop. The refrigerant loop works in the chiller to generate chilled water. The chilled water circulates in the chilled water loop connecting the chiller and the cooling coil held in the AHU. The chilled water flows in the cooling coil to absorb heat from the air passing though the coil. In the chiller, the chilled water transfers the heat to the refrigerant and the refrigerant transfers the heat to the cooling water flowing in and out of the chiller. The heat finally is discarded outdoors by the cooling water in a cooling tower. Chiller water can be generated by an electric chiller or an absorption chiller. One major difference between the two chillers is that the electric chiller uses electricity to drive the compressor and consumes a large amount of electricity, while the absorption chiller does not. Figure 4.12 shows a schematic of the central air conditioning system of the chilled water style. In the figure, it shows the electric chiller can be replaced conveniently by the absorption chiller to save electricity.

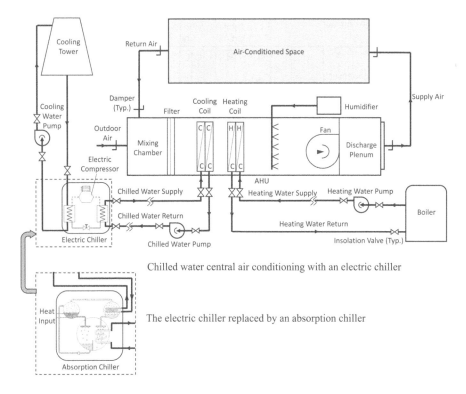

Chilled water central air conditioning with an electric chiller

The electric chiller replaced by an absorption chiller

FIGURE 4.12 A schematic of the chilled water central air conditioning.

4.4.3.2 Direct-Expansion Style

Direct-expansion (DX) systems are popular in central air conditioning. The term "direct" refers to the indoor air in cooling transferring heat directly to the refrigerant flowing inside the coil and the term "expansion" refers to the refrigerant expanding after the expansion valve in the system. The heat released from the indoor air is discharged from the refrigerant to the ambient air by an air-cold or water-cold condenser. A rooftop air handling unit (RTU) is a typical air conditioning system of DX style. Figure 4.13 shows schematics of an RTU with the air-cooled condenser and a DX air conditioning system with the water-cooled condenser, respectively.

4.4.4 STANDALONE AIR CONDITIONERS

4.4.4.1 Window A/C Unit

A window A/C unit is one of the simplest types of air conditioner. It is commonly used for cooling indoor air in one room or a limited area. The unit is a single box or case containing all the air conditioning components, including an evaporator coil, a condenser coil, a fan and a compressor. Typically, the compressor is the air-cold style. The window A/C unit operates on the working principle as described in Section 4.4.1. Figure 4.14 shows a schematic of a typical window A/C unit.

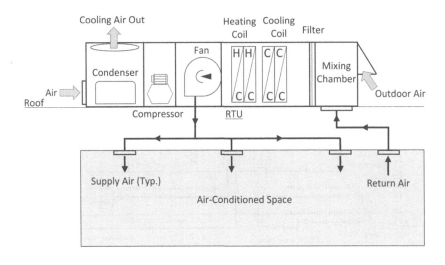

(a)

(b)

FIGURE 4.13 Schematics of the direct-expansion systems. (a) An RTU with the air-cooled condenser. (b) A DX air conditioning system with the water-cooled condenser.

4.4.4.2 Portable A/C Unit

A portable A/C unit, also known as a stand-up air conditioner, is a self-contained system. It is similar to the window A/C unit. However, the portable A/C unit is designed to stand on the floor instead of hanging on the window. The unit works to circulate the air in the room by taking the room air in and blowing cold air back to the room. Typically, the compressor is an air-cold style. The air after cooling the condenser is vented outdoors through an exhaust hose. Figure 4.15 shows a schematic of a typical portable A/C unit.

4.4.4.3 Ductless Air Conditioner

A ductless air conditioner is a mini-split air conditioning system, also known as a ductless mini-split air conditioner, which operates on the working principle as

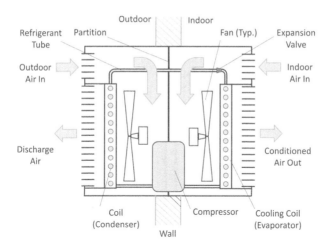

FIGURE 4.14   A schematic of a typical window A/C unit.

described in Section 4.4.2. The conditioner consists of an indoor unit and an out-door unit. The indoor unit houses an evaporator coil, an electric resistive heater, and a blower. The outdoor unit houses a condenser, a compressor, and a fan. The indoor unit is mounted on the wall without a ductwork connection. The unit circulates the room air in and blows out the conditioned air. The ductless air conditioner can operate in cooling mode and heating mode. In cooling mode, the room air passing through the evaporator coil becomes the cold air discharged back to the room. The heat taken from the room air is discarded to the ambient air

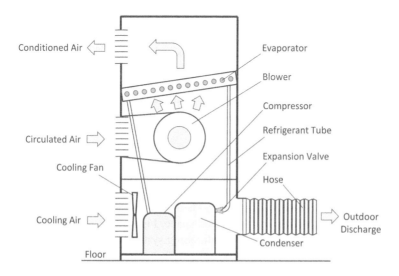

FIGURE 4.15   A schematic of a typical portable A/C unit.

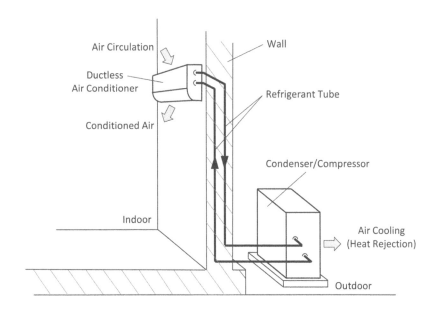

FIGURE 4.16   A schematic of a typical ductless air conditioner.

through the outdoor unit. In heating mode, the room air passing through the electrical resistive heater becomes the warm air discharged back to the room. The indoor and outdoor units are connected with copper tubing to circulate the refrigerant. The ductless air conditioner usually is designed to provide conditioned air in one specified area. Figure 4.16 shows a schematic of a typical ductless air conditioner.

As with the split system, a ductless air conditioner can work with a heat pump for heating and cooling. In hot weather, heat taken from the room is discarded outside through the heat pump coils. In cold weather, heat taken from the outside through the heat pump coils to warm the room air using the electric resistor.

4.4.5   APPLICATION IN DUCTWORK

4.4.5.1   Fan Coil Unit Application

A fan coil unit (FCU) generally consists of a cooling coil with chilled water, a heating coil with hot water or steam, and a fan to blow the conditioned air to the air-conditioned space. The coils and fan are housed in one box. The FCU works as a supplement air conditioner in ductwork. Therefore, the FCU is usually installed in the ductwork closer to the air-conditioned space. Figure 4.17 shows a schematic of the FCU in a multi-zone central air conditioning system.

The FCU system is suitable in applications where there are multiple small spaces requiring individual air conditioning control. It is widely used in multi-room buildings, such as hotels, office blocks, and schools.

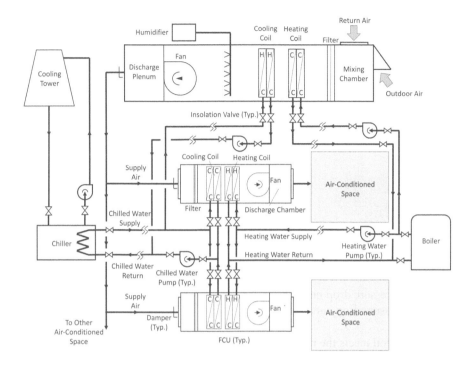

FIGURE 4.17 A schematic of the fan coil units in a multi-zone central air conditioning system.

4.4.5.2 Duct Coil

A duct coil, also known as a duct booster coil, is typically installed in the last ductwork near the air-conditioned space. The duct heating coil is also called the reheat coil. The purpose of using the duct coil is similar to the FCU to maintain the temperature level in the air-conditioned space. In application, normally there are upstream coils, such as in the AHU, that regulate the air temperature to an intermediate level and the duct coil adjusts the intermediate air temperature to the designed room temperature. Both the heating and cooling coils can be installed on one duct. Commonly, chilled water is used for cooling and hot water or steam is used for heating. Figure 4.18 shows a schematic of the duct coil application for heating and cooling.

4.5 COIL SELECTION

Heat and cooling coils are integral parts in an air conditioning system and a coil fluid piping system. The coils are critical components to the ultimate condition of the air-conditioned space. The selection of a coil in application requires applying the knowledge of thermodynamics, fluid mechanics, heat transfer, and also coil construction. A properly selected coil must meet three major requirements:

- Heat transfer load of the coil.

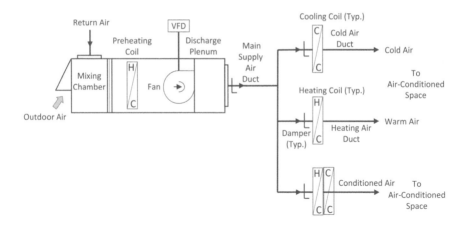

FIGURE 4.18 A schematic of the duct coil application.

- Fluid pressure drop inside the coil.
- Air pressure drop outside the coil.

If a selected coil meets the requirements of the heat transfer load of the coil and the pressure drop of the chilled water flowing inside the coil, but the coil has a higher airside pressure drop than specified or expected, the coil should be reselected since if the coil pressure resistance on the airside is too high, the fan may consume a higher power than that specified or the coil may not function properly with the existing fan.

4.5.1 HEATING COIL

The procedure of selection for a heating coil is illustrated in the following example.

Example 4.3

(Adapted from Marlo Heat Transfer Solutions, 2008.)

A hot water heating coil needs to be selected for a heating system by the specifications:

Flow rate $\dot{V}$ of air entering the coil, m³/min	85
T_{db} and RH of air entering the coil, °C and %	4.5 and 60
T_{db} of desired air leaving the coil, °C	32
Temperature of heating water entering the coil, °C	82
Temperature of desired heating water leaving the coil °C	71
Coil dimension, height × length, m	0.305 × 1.524
Material of coil tube and fin	Copper
Diameter of coil tube, mm	15.875

Referring to Table(A) *Coil Data* and Figure Examples 4.3 and 4.4 Coil configuration for the coil data and configuration.

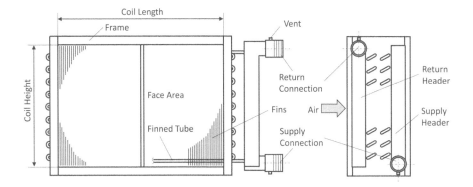

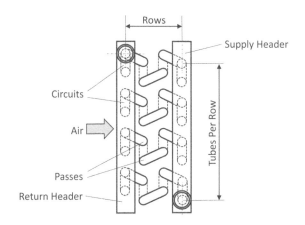

A typical Coil of 4 Rows, 4 Circuits, 8 tubes per Row, and 8 Passes

- Rows: number of tubes along airflow direction horizontally
- Tubes per row: number of tubes facing airflow vertically
- Circuits: number of tubes connecting supply header to return header for fluid flow
- Passes: number of tubes across the coil face area per circuit

FIGURE Examples 4.3 and 4.4 Coil configuration.

SOLUTION

1. Determine coil size.
   Referring to Table(B) *Coil Face Area (m²)*, a coil size of 0.305 m and 1.524 m (height and width) with 8 tubes per row is selected. The face area is 0.465 m².
2. Determine face velocity of air.

$$V = \frac{\dot{V}}{A_{face}} = \frac{85\ \dfrac{m^3}{min}}{0.465\ m^2} = 182.8\ \frac{m}{min}$$

3. Calculate the rate of heat transfer.

The state of the air entering the coil is specified by the given T_{db} and relative humidity (RH). Referring to Appendix A.4 Psychrometric Chart at a Pressure of 1 atm (101.325 kPa), the specific volume of the air entering the coil is

$$v = 0.79 \frac{m^3}{kg}$$

From the chart, the enthalpies of the entering and exiting air are

$$h_{in} = 12.8 \frac{kJ}{kg \ dry \ air} \quad h_{out} = 40.3 \frac{kJ}{kg \ dry \ air}$$

Using Equation (2.3b) (see Chapter 2), therefore, the required air heating load (sensible heating) is determined,

$$\dot{Q} = \dot{m}(h_{out} - h_{in}) = \frac{85 \frac{m^3}{min}}{0.79 \frac{m^3}{kg}} (40.3 - 12.8) \frac{kJ}{kg \ dry \ air}$$

$$= 2,958.86 \frac{kJ}{min} = \left(2,958.86 \frac{kJ}{min}\right)\left(60 \frac{min}{hr}\right)$$

$$= 177,531.65 \frac{kJ}{hr} = 177,531.65 \frac{kJ}{hr}\left(\frac{1,000 \frac{J}{kJ}}{3,600 \frac{s}{hr}}\right)$$

$$= 49,314.4 \frac{J}{s} = 49,314.4 \ W$$

4. Find the required flow rate of the heating water.

Using Equation (2.2) (see Chapter 2), the required mass flow rate of the heating water is found,

$$\dot{m} = \frac{\dot{Q}}{c_p(T_{out} - T_{in})} = \frac{2,958.86 \frac{kJ}{min}}{4.1942 \frac{kJ}{kg.K}(82 - 71)°C} = 64.13 \frac{kg}{min}$$

and the required volume flow rate is

$$\dot{V} = \frac{\dot{m}}{\rho} = \frac{64.13 \frac{kg}{min}}{973.83 \frac{kg}{m^3}} = 0.0659 \frac{m^3}{min}$$

where $c_p = 4.1942$ kJ/kg.K and $\rho = 973.83$ kg/m^3 are determined by the average water temperature $(82°C + 71°C)/2 = 76.5°C$.

5. Select the desired serpentine and row circuits to determine the heating water velocity.

- Recommend water velocity from 0.61 to 1.82 m/s to achieve turbulent flow velocity and reduce tube erosion.
- Choose the serpentine defined as the ratio of number of rows to number of passes.

$$\text{Serpentine} = \frac{\text{Rows}}{\text{Passes}}$$

The serpentine of the heating coil is selected to be 0.5.

- Referring to Table(C) *Number of Circuits*, the coil with 2 rows and 8 facing tubes will have 4 circuits. If it is necessary, the selected serpentine can be changed to keep the water velocity inside the coil tube within the range 0.61 to 1.82 m/s.
- Referring to Table(D) *Tube Wall Thickness Factors*, the factor is 99.5 from the coil tube data. The heating water velocity in the coil tube, therefore, is

$$V_w = \frac{\dot{V} \times \text{Factor}}{\text{No. of circuits}}$$

$$= \frac{\left(0.0659 \, \dfrac{m^3}{min}\right) 99.5}{4} = 1.6393 \, \frac{m}{s}$$

which is within the recommended range of 0.61 to 1.82 m/s.

6. Determine LMTD.

Knowing the temperature differences,

$$\Delta T_1 = T_{h,in} - T_{c,out} = (82 - 32)°C = 50°C$$

$$\Delta T_2 = T_{h,out} - T_{c,in} = (71 - 4.5)°C = 66.5°C$$

and using Equation (2.70) (see Chapter 2), the LMTD of counter flow is

$$\Delta T_{lm,cf} = \frac{\Delta T_1 - \Delta T_2}{\ln(\Delta T_1)/(\Delta 2)} = \frac{(50 - 66.5)°C}{\ln(50°C)/(66.5°C)} = 57.86°C$$

For the staggered cross flow, the correction factor F is 0.98 based on the temperatures of the fluids (air and heating water) entering and exiting the coil. Using Equation (4.2), the LMTD of the staggered cross flow is found,

$$\Delta T_{eq} = F\Delta T_{lm,cf} = 0.98(57.86°C) = 56.70°C$$

7. Determine the overall heat transfer coefficient U of the coil
    Referring to Table(F) U of Multi-Row (W/m². °C), for a multi-row coil at the air face velocity 182.8 m/min and the water velocity 1.6393 m/s, U is found to be
    1,086.5 W/m².°C for 6 fins per 25.4 mm
    or
    1,331.6 W/m².°C for 8 fins per 25.4 mm
8. Determine number of rows and fin spacing by using the formula,

$$Rows = \frac{\dot{Q}}{A_{face} \times \Delta T_{eq} \ U}$$

For 6 fins per 2.54 cm,

$$Rows = \frac{49{,}314.4 \ W}{\left(0.465 \ m^2\right)\left(56.70°C\right)\left(1{,}086.5 \ \dfrac{W}{m^2°C}\right)}$$

$$= 1.72 \ Rows$$

For 8 fins per 2.54 cm,

$$Rows = \frac{49{,}314.4 \ W}{\left(0.465 \ m^2\right)\left(56.70°C\right)\left(1{,}331.6 \ \dfrac{W}{m^2°C}\right)}$$

$$= 1.41 \ Rows$$

The selection of a 2-row coil with 6 fins per 25.4 mm is more practical.
9. Estimate the air pressure drop ΔP_a for the coil of 2 rows and 6 fins per 25.4 mm.
    • Referring to Table(G) Dry Air Pressure Drop (mmH₂O) Based on Row Arrangement, at face velocity of 182.8 m/min, the air pressure drop is

$$\Delta P_a = \frac{Air \ pressure \ drop}{row} \times Rows$$

$$\Delta P_a = \left(3.8617 \ \frac{mm \ H_2O}{row}\right)\left(2 \ row\right) = 7.7234 \ mm \ H_2O$$

10. Estimate water pressure drop ΔP_w.
    • Referring to Table(J) Number of Passes per Circuit, for the coil with 2 row, 0.5 serpentine, there are 4 passes in the coil. The total length of one water circuit is

$$Circuit \ Length = \left(length \times pass\right) + \left(Ll_{elbow}\right)\left(Passes - 1\right)$$

$$= \left(1.524 \ m\right)4 + \left(0.3788 \ m\right)\left(4 - 1\right) = 7.23 \ m$$

where Ll$_{elbow}$ is linear length of an elbow. Referring to Table(K) *Water Pressure Drop* at 1.6393 m/s, the pressure drops of the tube and head are

$$\Delta P_{tube} = 0.2239 \text{ m/unit meter}$$

$$\Delta P_{header} = 1.6446 \text{ m/unit meter}$$

- Referring to Table(l) *Water Temperature Correction Factor,* an average water temperature of 76.5°C, F1 = 0.7957 and F2 = 1.0549 are found. Therefore, the total pressure drop of water through the coil is determined to be

$$\Delta P_w = \text{Circuited Length} \times \Delta P_{tube} \times F1 + \Delta P_{header} \times F2$$

$$= (7.23 \text{ m})\left(0.2239 \frac{m}{\text{unit meter}}\right)0.7957 + (1.6446 \text{ m})\,1.0549$$

$$= 3.023 \text{ m H}_2\text{O}$$

4.5.2 COOLING COIL

The procedure of selection for a cooling coil is illustrated in the following example.

Example 4.4

(Adapted from Marlo Heat Transfer Solutions, 2008)

A cooling coil needs to be selected in a cooling with dehumidifying process by these specifications:

Volume flow rate $\dot{V}$ of air entering the coil, m³/min	140.0
T_{db} of air entering the coil, °C	26.7
T_{wb} of air entering the coil, °C	19.4
T_{db} of desired air leaving the coil, °C	11.7
T_{wb} of desired air leaving the coil, °C	11.4
Temperature of CHW entering the coil, °C	7.2
Temperature of CHW leaving the coil, °C	12.8
Maxim allowable pressure drop of CHW, m H$_2$0	4.85
Coil Dimension, height × length, m	0.610 × 1.524
Material of coil tube and fin	Copper
Diameter of coil tube, mm	15.875

Referring to Table(A) *Coil Data* and Figure Examples 4.3 and 4.4 Coil structure for the coil data and configuration.

SOLUTION

1. Determine Coil Size.

Referring to Table(B) *Coil Face Area*, a coil size of 0.610 m and 1.524 m (height and width) with 16 tubes per row is selected. The face area is 0.929 m².

2. Determine face velocity of air.

$$V = \frac{\dot{V}}{A_{face}} = \frac{140 \, \frac{m^3}{min}}{0.929 \, m^2} = 150.7 \, \frac{m}{min}$$

3. Calculate the rate of heat transfer.

The state of the air entering the coil is specified by the given T_{db} and T_{wb}. Referring to Appendix A.4 Psychrometric Chart at a Pressure of 1 atm (101.325 kPa), the specific volume of the air entering the coil is,

$$v = 0.8645 \, \frac{m^3}{kg}$$

From the chart, the enthalpies of the entering and exiting air are found,

$$h_{out} = 32.8 \, \frac{kJ}{kg \, dry \, air}$$

$$h_{in} = 55.6 \, \frac{kJ}{kg \, dry \, air}$$

Using Equation (2.3b) (see Chapter 2), therefore, the required air load is determined to be

$$\dot{Q} = \dot{m}\left(h_{out} - h_{in}\right) = \frac{140.0 \, \frac{m^3}{min}}{0.8645 \, \frac{m^3}{kg}} \left(55.6 - 32.8\right) \frac{kJ}{kg \, dry \, air}$$

$$= 3,692.31 \, \frac{kJ}{min} = \left(3,692.31 \, \frac{kJ}{min}\right)\left(60 \, \frac{min}{hr}\right)$$

$$= 221,538.6 \, \frac{kJ}{hr} = 221,538.6 \, \frac{kJ}{hr} \left(\frac{1,000 \, \frac{J}{kJ}}{3,600 \, \frac{s}{hr}}\right)$$

$$= 61,538.5 \, \frac{J}{s} = 61,538.5 \, W = 61.54 \, kW$$

4. Determine the rate of heat transfer per unit area of the face area.

$$UHTR = \frac{\dot{Q}}{A_{face}} = \frac{61,538.5 \, W}{\left(0.929 \, m^2\right)\left(1,000 \, \frac{W}{kW}\right)} = 66.24 \, \frac{kW}{m^2}$$

5. Find the required flow rate of the CHW.

Using Equation (2.2) (see Chapter 2), the required mass flow rate of the CHW is found as

$$\dot{m}_w = \frac{\dot{Q}}{c_p\left(T_{out} - T_{in}\right)} = \frac{3{,}692.31\ \dfrac{kJ}{min}}{4.194\ \dfrac{kJ}{kg.K}\left(12.8 - 7.2\right)°C} = 157.2\ \frac{kg}{min}$$

and the required volume flow rate is

$$\dot{V}_w = \frac{\dot{m}}{\rho} = \frac{157.2\ \dfrac{kg}{min}}{999.7\ \dfrac{kg}{m^3}} = 0.1573\ \frac{m^3}{min}$$

where $c_p = 4.194$ kJ/kg.K and $\rho = 999.7$ kg/m³ are determined by the average water temperature (12.8°C + 7.2°C)/2 = 10°C.

6. Select the desired serpentine and row circuits to determine the CHW velocity.
   - Recommend water velocity from 0.61 to 1.82 m/s (2 to 6 ft/s) to achieve turbulent flow velocity and reduce tube erosion.
   - Choose the serpentine defined as the ratio of number of rows to number of passes:

$$\text{Serpentine} = \frac{\text{Rows}}{\text{Passes}}$$

   Serpentine of the cooling coil is selected to be 1.0.
   - Referring to Table(C) *Number of Circuits*, the coil with 6 rows and 16 facing tubes will have 16 circuits. If it is necessary, the selected serpentine can be changed to keep the water velocity inside the coil tube within the range 0.61 to 1.82 m/s.
   - Referring to Table(D) *Tube Wall Thickness Factors*, the factor is 99.5 from the coil tube data. The CHW velocity in the coil tube, therefore, is

$$V_w = \frac{\dot{V}_w \times \text{Factor}}{\text{No. of circuits}} = \frac{\left(0.1573\ \dfrac{m^3}{min}\right)99.5}{16} = 0.9782\ \frac{m}{s}$$

   which is within the range of 0.61 to 1.82 m/s.

7. Select number of rows and find spacing.

Referring to Table(E) *Selection of Number of Rows and Fins* with
   - Air entering the coil at 26.7°C T_{db} and 19.4°C T_{wb}
   - Air face velocity 150.7 $\frac{m}{min}$
   - CHW entering the coil at 7.2°C
   - CHW entering the coil at 12.8°C
   - CHW velocity 0.9782 $\frac{m}{s}$

   A coil of 6 rows and 12 fins per 25.4 mm is selected. The coil can have
   - Unit heat transfer rate of 71.34 kW/m²

- Exiting air at 11.28°C T_{db}/11.18°C T_{wb}
  The total heat capacity is

$$\dot{Q}_{coil} = UHTR \times FA = \left(71.34 \ \frac{kW}{m^2}\right)\left(0.929 \ m^2\right)$$

$$= 66.28 \ kW$$

and the sensible heat is

$$\dot{Q}_{sensible} = \dot{m}c_p\left(T_i - T_o\right) = \left(\frac{140.0 \ \dfrac{m^3}{min}}{0.8645 \ \dfrac{m^3}{kg}}\right)\left(1.006 \ \frac{kJ}{kg.K}\right)\left(26.7 - 11.7\right)°C$$

$$= 2,441.3 \ \frac{kJ}{min} = \frac{911.42 \ \dfrac{kJ}{min}}{60 \ \dfrac{s}{min}} = 40.69 \ kW$$

where $c_p = 1.006$ kJ/kg.K from Appendix A.3 Specific Heat of Air.
8. Determine the sensible heat ratio (SHR).

$$SHR = \frac{Sensible \ Heat}{Total \ Heat} = \frac{40.69 \ kW}{66.28 \ kW} = 0.6149 = 61.49\%$$

9. Estimate air pressure drop ΔP_a for the coil of 6 rows and 12 fins per 25.4 mm.
   - Referring to Table(H) *Wet Air Pressure Drop (mmH$_2$O) Based on Row Arrangement*, at 150.7 m/min face velocity and 12 fins per 25.4 mm, the air pressure drop per row is 6.80 mm H$_2$O.
   - Referring to Table(I) *Air Pressure Drop Correction*, with SHR 0.62, the air pressure correction factor is 1.0, then,

$$\Delta P_a = \frac{Air \ pressure \ drop}{row} \times Rows \times \left(Air \ pressure \ correction\right)$$

$$= \left(6.80 \ mm \ H_2O\right)\left(6 \ rows\right)1.0 = 40.8 \ mm \ H_2O$$

10. Estimate CHW pressure drop ΔP_w.
    - Referring to Table(J) *Number of Passes per Circuit* for the coil with 6 row, 1 serpentine and 6 passes, the total length of one water circuit is

$$Circuit \ Length = \left(length \times 6\right) + \left(Ll_{elbow}\right)\left(6 - 1\right)$$

$$= \left(1.524 \ m\right)6 + \left(0.3788 \ m\right)\left(6 - 1\right) = 11.04 \ m$$

where Ll_{elbow} is linear length of an elbow. Referring to Table(k) *Water Pressure Drop*, at 0.9782 m/s, the pressure drops of tube and head are

$$\Delta P_{tube} = 0.1044 \text{ m/unit meter}$$

$$\Delta P_{header} = 2.061 \text{ m/unit meter}$$

- Referring to Table(L) *Water Temperature Correction Factor* at average water temperature of 10°F, F1 = 1.04 and F2 = 0.99 are found. Therefore, the total pressure of water inside the coil is determined to be

$$\Delta P_w = \text{Circuited Length} \times \Delta P_{tube} \times F1 + \Delta P_{header} \times F2$$

$$= (11.04 \text{ m})\left(0.1044 \; \frac{m}{\text{unit meter}}\right)1.04 + (2.061 \text{ m}) \, 0.99$$

$$= 3.239 \text{ m H}_2\text{O}$$

TABLE(A)
Coil Data

Tubes	Fins	Headers
Diameter: 15.875 mm	Type: Plate	Material: Copper
Wall thickness: 0.635 mm	Thickness: 0.254 mm	
Material: Copper	Material: Copper	

TABLE(B)
Coil Face Area (m²)

Finned Height (m)	Tubes Per Row	Finned Length (m)												
		0.457	0.533	0.610	0.762	0.914	1.067	1.219	1.372	1.524	1.676	1.829	1.981	2.134	2.286
0.152	4	0.070	0.081	0.093	0.116	0.139	0.163	0.186	0.209	0.232	0.255	0.279	0.302	0.325	0.348
0.229	6	0.105	0.122	0.139	0.174	0.209	0.244	0.279	0.314	0.348	0.383	0.418	0.453	0.488	0.523
0.305	8	0.139	0.163	0.186	0.232	0.279	0.325	0.372	0.418	0.465	0.511	0.557	0.604	0.650	0.697
0.381	10	0.174	0.203	0.232	0.290	0.348	0.406	0.465	0.523	0.581	0.639	0.697	0.755	0.813	0.871
0.457	12	0.209	0.244	0.279	0.348	0.418	0.488	0.557	0.627	0.697	0.766	0.836	0.906	0.975	1.045
0.533	14	0.244	0.285	0.325	0.406	0.488	0.569	0.650	0.732	0.813	0.894	0.975	1.057	1.138	1.219
0.610	16	0.279	0.325	0.372	0.465	0.557	0.650	0.743	0.836	0.929	1.022	1.115	1.208	1.301	1.394
0.686	18	0.314	0.366	0.418	0.523	0.627	0.732	0.836	0.941	1.045	1.150	1.254	1.359	1.463	1.568
0.762	20	0.348	0.406	0.465	0.581	0.697	0.813	0.929	1.045	1.161	1.277	1.394	1.510	1.626	1.742
0.838	22	0.383	0.447	0.511	0.639	0.766	0.894	1.022	1.150	1.277	1.405	1.533	1.661	1.788	1.916

TABLE(C)
Number of Row Circuits

	Serpentine					
	0.5	0.75	1	1.25	1.5	2
	Rows					
Tubes in Face (TF)	1, 2, 3, 4, 5, 6, 8, 10	3,6,8	1, 2, 3, 4, 5, 6, 8, 10	5,10	3, 6	2, 4, 6, 8, 10
4	2	3	4	5	6	8
5			5			10
6	3		6		9	12
7			7			14
8	4	6	8	10	12	16
9			9			18
10	5		10		15	20
11			11			22
12						
12	6	9	12	15	18	24
13			13			26
14	7		14		21	28
15			15			30
16	8	12	16	20	24	32
17			17			34
18	9		18		27	36
19			19			38
20	10		20	25	30	40

TABLE(D)
Tube Wall Thickness Factors

Wall Thickness (mm)	Factor
0.6350	99.5
0.7112	101.6
0.8890	106.8
1.2446	118.4

TABLE(E)
Selection of Number of Rows and Fins

CHW (m/s)	Fins per 25.4 mm	Row	121.2 UHTR (kW/m²)	121.2 Leaving DBT (°C)	121.2 Leaving WBT (°C)	151.5 UHTR (kW/m²)	151.5 Leaving DBT (°C)	151.5 Leaving WBT (°C)	181.8 UHTR (kW/m²)	181.8 Leaving DBT (°C)	181.8 Leaving WBT (°C)
0.61	8	4	41.0	14.4	13.8	45.7	15.2	14.5	15.8	15.8	15.0
	10		44.5	13.6	13.3	49.5	14.4	14.1	17.1	15.1	14.6
	12		47.0	13.1	12.9	52.4	13.9	13.7	57.1	14.6	14.3
	14		48.9	12.8	12.6	54.6	13.6	13.4	59.6	14.2	14.0
	8	6	52.4	12.3	12.1	59.0	13.1	12.8	65.3	13.8	13.4
	10		55.2	11.7	11.6	63.1	12.5	12.3	69.7	13.2	13.0
	12		57.7	11.3	11.1	65.9	12.1	11.9	72.9	12.8	12.7
	14		59.6	10.9	10.8	67.8	11.8	11.7	75.4	12.5	12.4
	8	8	60.3	10.9	10.7	69.1	11.7	11.5	77.0	12.4	12.2
	10		62.8	10.4	10.2	72.9	11.2	11.1	81.4	11.9	11.7
	12		65.0	10.0	9.9	75.4	10.8	10.7	84.2	11.6	11.4
	14		66.6	9.7	9.6	77.3	10.6	10.4	86.8	11.3	11.1
0.91	8	4	44.5	13.8	13.3	50.2	14.7	13.9	54.6	15.3	14.5
	10		48.3	13.0	12.7	54.6	13.8	13.4	59.6	14.5	14.0
	12		51.1	12.4	12.2	57.7	13.3	13.0	63.4	13.9	13.6
	14		53.3	12.0	11.8	60.6	12.8	12.7	66.6	13.5	13.3
	8	6	55.8	11.7	11.4	64.0	12.5	12.2	71.0	13.2	12.8
	10		59.3	11.0	10.8	68.5	11.8	11.6	76.0	12.5	12.3
	12		62.1	10.5	10.4	71.6	11.3	11.2	79.8	12.1	11.9
	14		64.0	10.2	10.0	74.1	11.0	10.8	82.6	11.7	11.6
	8	8	63.4	10.3	10.1	73.8	11.1	10.9	83.0	11.7	11.5
	10		66.6	9.7	9.6	77.9	10.5	10.3	88.0	11.2	11.0
	12		68.8	9.3	9.2	80.8	10.1	9.9	91.2	10.7	10.6
	14		70.3	9.0	8.9	83.0	9.8	9.6	94.0	10.4	10.3
1.12	8	4	46.7	13.5	12.9	52.7	14.3	13.6	58.0	15.0	14.2
	10		50.8	12.6	12.3	57.4	13.4	13.1	63.4	14.1	13.6
	12		53.9	11.9	11.8	61.2	12.8	12.6	67.8	13.5	13.2
	14		56.5	11.5	11.3	64.4	12.3	12.2	71.0	13.1	12.8
	8	6	58.0	11.3	11.1	66.9	12.1	11.8	74.8	12.8	12.4
	10		61.8	10.6	10.4	71.6	11.4	11.2	80.1	12.1	11.8
	12		64.7	10.1	9.9	75.1	10.9	10.7	84.2	11.6	11.4
	14		66.9	9.7	9.6	77.9	10.5	10.3	87.4	11.2	11.0
	8	8	65.3	9.9	9.8	76.7	10.7	10.5	86.8	11.3	11.1
	10		68.8	9.3	9.2	80.8	10.1	9.9	91.8	10.7	10.6
	12		71.0	8.9	8.8	83.9	9.6	9.4	95.6	10.2	10.1
	14		72.6	8.6	8.5	86.1	9.3	9.2	98.4	9.9	9.8

UHTR: Unit Heat Transfer rate
DBT: Dry-Bulb Temperature
WBT: Wet-Bulb Temperature

TABLE(F)
U of Multi-Row (W/m². °C)

Fins Per 25.4 mm	V_{water} (m/s)	Face Velocity V (m/min)					
		60	121	182	242	303	364
6	0.61	623.4	840.9	984.0	1,090.7	1,175.9	1,246.3
	0.91	642.7	876.7	1,033.4	1,152.1	1,247.4	1,326.9
	1.21	654.1	897.1	1,062.3	1,187.8	1,289.5	1,374.6
	1.52	660.9	910.7	1,081.1	1,211.7	1,317.9	1,407.0
	1.82	666.0	920.4	1,094.7	1,228.7	1,337.7	1,429.7
	2.42	672.8	933.5	1,112.9	1,252.0	1,365.5	1,461.5
8	0.61	769.9	1,020.3	1,180.4	1,298.0	1,390.5	1,466.0
	0.91	800.0	1,073.7	1,252.6	1,385.4	1,491.6	1,579.0
	1.21	817.1	1,104.4	1,295.1	1,437.7	1,551.8	1,078.8
	1.52	828.4	1,125.4	1,323.5	1,472.9	1,592.7	1,693.2
	1.82	836.4	1,140.1	1,344.0	1,498.4	1,622.8	1,726.7
	2.42	847.1	1,160.0	1,371.2	1,532.5	1,663.1	1,772.7
10	0.61	902.8	1,177.6	1,347.9	1,474.0	1,570.5	1,648.9
	0.91	944.2	1,249.1	1,445.0	1,588.1	1,701.1	1,793.1
	1.21	968.1	1,291.2	1,501.8	1,656.8	1,780.0	1,881.7
	1.52	984.0	1,319.6	1,539.9	1,704.0	1,834.0	1,941.9
	1.82	995.3	1,340.0	1,567.7	1,737.5	1,873.7	1,981.6
	2.42	1,010.1	1,367.2	1,605.2	1,784.0	1,927.7	2,047.5

TABLE(G)
Dry Air Pressure Drop (mm H₂O) Based on Row Arrangement

	Fins Per 25.4 mm	Face Velocity (m/min)										
		60	90	120	150	180	210	240	270	300	330	360
One Row	6	0.81	1.52	2.39	3.40	4.50	5.74	7.06	8.48	9.98	11.58	13.28
	8	0.99	1.78	2.79	3.94	5.23	6.65	8.18	9.80	11.56	13.39	15.34
	10	1.09	2.03	3.18	4.50	5.97	7.57	9.30	11.15	13.13	15.21	17.40
	12	1.24	2.31	3.58	5.05	6.68	8.48	10.41	12.50	14.68	17.02	19.46
	14	1.37	2.57	3.99	5.61	7.42	9.40	11.53	13.82	16.26	18.82	21.51
Two or More Rows	6	0.58	1.17	1.88	2.74	3.76	4.85	6.10	7.44	8.89	10.44	12.09
	8	0.74	1.42	2.29	3.33	4.50	5.79	7.21	8.76	10.44	12.22	14.12
	10	0.86	1.70	2.69	3.89	5.23	6.71	8.36	10.11	11.99	14.02	16.15
	12	1.02	1.96	3.10	4.45	5.97	7.65	9.47	11.46	13.56	15.80	18.16
	14	1.17	2.24	3.53	5.03	6.71	8.56	10.59	12.78	15.11	17.58	20.19

TABLE(H)
Wet Air Pressure Drop (mm H₂O) Based on Row Arrangement

	Fins Per 25.4 mm	Face Velocity (m/min)										
		61	76	91	106	121	136	152	167	182	197	212
Two or	6	0.94	1.35	1.80	2.29	2.82	3.40	4.01	4.67	5.36	6.07	6.81
More	8	1.27	1.78	2.34	2.92	3.56	4.24	4.98	5.74	6.53	7.37	8.23
Rows	10	1.60	2.18	2.84	3.56	4.32	5.11	5.94	6.81	7.72	8.66	9.63
	12	1.91	2.62	3.38	4.19	5.05	5.94	6.88	7.87	9.14	9.96	11.05
	14	2.24	3.02	3.89	4.83	5.79	6.81	7.85	8.94	10.08	11.25	12.45

TABLE(I)
Air Pressure Drop Correction Factors

SHR	APF
0.64 or less	1
0.065 – 0.74	0.9
0.75 – 0.84	0.8
0.85 – 0.94	0.7

APT: Air Pressure Drop Correction Factor.

TABLE(J)
Number of Passes Per Circuit

	Rows								
Serpentine	1	2	3	4	5	6	8	10	12
1/2	2	4	6	8	10	12	16	20	24
3/4	—	—	4	—	—	8	—	—	16*
1	*1	2	*3	4	*5	6	8	10	12
1 1/2	—	—	2	—	—	4	—	—	8
2	—	1	—	2	—	3	4	*5	6

* Opposite end connection

TABLE(K)
Water Pressure Drop

V_w (m/s)	Tube Pres. Drop (m/unit m)	Header Pres. Drop (m)	V_w (m/s)	Tube Pres. Drop (m/unit m)	Header Pres. Drop (m)	V_w (m/s)	Tube Pres. Drop (m/unit m)	Header Pres. Drop (m)	V_w (m/s)	Tube Pres. Drop (m/unit m)	Header Pres. Drop (m)
0.61	0.052	0.820	1.06	0.118	0.730	1.52	0.200	1.412	1.97	0.296	2.338
0.64	0.056	0.915	1.09	0.122	0.769	1.55	0.206	1.471	2.00	0.303	2.406
0.67	0.060	1.010	1.12	0.126	0.808	1.58	0.212	1.529	2.03	0.310	2.474
0.70	0.065	1.105	1.15	0.131	0.847	1.61	0.218	1.588	2.06	0.317	2.542
0.73	0.069	1.200	1.18	0.135	0.885	1.64	0.224	1.646	2.09	0.324	2.609
0.76	0.073	1.295	1.21	0.140	0.924	1.67	0.230	1.705	2.12	0.331	2.679
0.79	0.077	1.390	1.24	0.146	0.973	1.70	0.236	1.763	2.15	0.338	2.756
0.82	0.081	1.485	1.27	0.152	1.022	1.73	0.242	1.822	2.18	0.345	2.833
0.85	0.086	1.580	1.30	0.158	1.071	1.76	0.248	1.880	2.21	0.352	2.911
0.88	0.090	1.675	1.33	0.164	1.119	1.79	0.254	1.938	2.24	0.359	2.988
0.91	0.094	1.770	1.36	0.170	1.168	1.82	0.260	1.997	2.27	0.367	3.065
0.94	0.099	1.898	1.39	0.176	1.217	1.85	0.267	2.065	2.30	0.374	3.142
0.97	0.103	2.026	1.42	0.182	1.266	1.88	0.274	2.133	2.33	0.381	3.220
1.00	0.108	2.154	1.45	0.188	1.315	1.91	0.281	2.202	2.36	0.388	3.297
1.03	0.112	2.282	1.48	0.194	1.363	1.94	0.288	2.270	2.39	0.395	3.374

TABLE(L)
Water Temperature Correction Factor

Average Water Temperature (°C)	Tube (F1)	Header (F2)	Average Water Temperature (°C)	Tube F1	Header F2
7.2	1.06	0.99	37.8	0.908	1.03
10.0	1.04	0.99	46.1	0.877	1.03
12.8	1.02	0.995	60.0	0.825	1.03
15.6	1	1	65.6	0.815	1.04
18.3	0.985	1.015	71.1	0.804	1.05
21.1	0.97	1.03	82.2	0.787	1.06
26.7	0.949	1.03	93.3	0.774	1.08

4.5.3 COMPUTER SOFTWARE

Computer software used for coil selection is a computer program based on thermo-dynamics, fluid mechanics, heat transfer, and coil construction. There are a variety of software titles for coil selection available for engineering applications. In general, the software functions are the same. Software applications help make coil selection quicker. In software selection, one needs to input the specifications and data into the program by following the software instructions. Figure 4.19 shows the computer screens of software for heating and cooling coil selection, respectively.

(a)

(b)

FIGURE 4.19 Marlo's computer software of coil selection. (a) Heating coil (hot water). (b) Cooling coil (chilled water). (Extracted from Marlo Heat Transfer Solutions, Coil & AHU Selection Software, 2022.)

5 Heating and Cooling Processes of Air Through Coils

5.1 INTRODUCTION

In engineering practice, a heating coil is for heating and a cooling coil is for cooling as air passes through the coil in air conditioning. Heating without water or steam spray is a simple heating process called a sensible heat transfer process. Cooling, however, may be a simple cooling process called a sensible cooling process or a process with dehumidifying. The process of cooling involving dehumidifying is complicated. The heating and cooling processes of air passing through coils can be graphically analyzed by the psychrometric chart.

5.2 PROCESS ANALYSIS

Simple heating and simple cooling processes of air passing through coils involves a temperature change of the air but the moisture content in the air stays constant during the process. This is shown by straight horizontal lines represented on the psychrometric chart as described in Section 3.3.1 (see Chapter 3). Cooling with dehumidifying, however, causes both temperature and moisture content to drop simultaneously. As air enters the coil, some of the air impinges onto the cooling coil tubes. The moisture in this part of the air is condensed since the surface temperature of the coil tubes is lower than the dew point of the air. The moisture content starts to decrease in the air. The part of the air without contact to the coil tubes experiences sensible cooling and its moisture content will reduce until the air impinges onto the following coil tubes. The process proceeds in the same pattern row by row as the air flows through the tube bank in the coil. The process line of cooling with dehumidifying on the psychrometric chart, therefore, is downward to the left. The line is not straight, but curves toward the saturation line. Figure 5.1(a) shows schematics of the process lines of simple heating, simple cooling, and cooling with dehumidifying of air passing through the coils. In the figure, line 1–2 represents a simple heating process, line 1–3 represents a simple cooling process, and line 1–4 represents a cooling with dehumidifying process. Figure 5.1(b) shows a cooling with dehumidifying process on the psychrometric chart. The curve shape of the process line on the chart depends on the coil bypass factor.

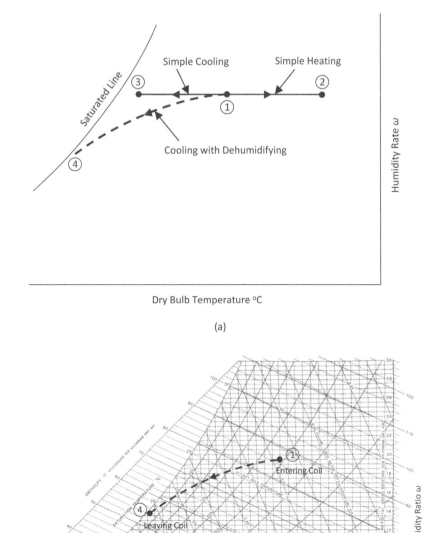

FIGURE 5.1 Process lines on the psychrometric chart. (a) Process lines of simple heating, simple cooling, and cooling with dehumidifying. (b) A cooling with dehumidifying process on the psychrometric chart.

5.3 FACTORS

Many factors can have impacts on the process of cooling with dehumidifying for air passing through the coil. Some factors have significant effects and directly affect the process. These factors can be graphically analyzed and represented on the psychrometric chart.

5.3.1 BYPASS FACTOR

The bypass factor (BF) is the ratio of the part of air passing through coil tubes without contact with coil tubes to the total quantity of air passing through the coil. Figure 5.2 is a schematic of the definition of the BF.

Alternatively, the BF can be represented by a contact factor (CF). The CF is the complementary of the BF, i.e.,

$$CF = 1 - BF \tag{5.1}$$

The BF of each coil row may not be the same. The BF affects the process line shape of the cooling with dehumidifying on the psychrometric chart. Figure 5.3 is a schematic of the BF effect to the process of air passing an 8-row cooling coil.

Determination of the BF in a coil is not an easy work. It involves coil configuration and airflow conditions, such as

1. Number of coil rows.
2. Pitch of the coil fins, i.e., density of fins.
3. Airflow types, turbulent or laminar.
4. Direction of airflow to the coil.

In general, the smaller the pitch, the higher number of rows, and the lower airflow velocity will cause the smaller BF. In engineering practice, the average BF in a coil is estimated from experience or from recommendation of the A/C industry. For applications with higher internal sensible loads or that require a large amount of outdoor air for ventilation, the BF is in a range of 0.05 to 0.1. For typical comfort applications, the BF is in a range of 0.1 to 0.2.

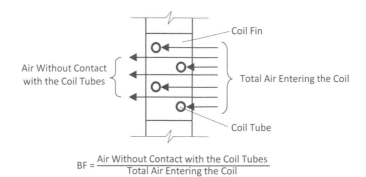

$$BF = \frac{\text{Air Without Contact with the Coil Tubes}}{\text{Total Air Entering the Coil}}$$

FIGURE 5.2 A schematic of definition of the bypass factor.

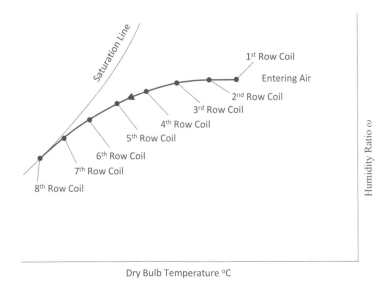

FIGURE 5.3 Bypass factor effect in the process of cooling with dehumidifying.

5.3.2 SENSIBLE HEAT FACTOR

Sensible heat factor (SHF), also called sensible heat ratio (SHR), is defined as the ratio of sensible heat to total heat, which is a summation of sensible and latent heat. The ratio is expressed as

$$\begin{aligned} SHF &= \frac{\text{Sensible Heat}}{\text{Total Heat}} \\ &= \frac{\text{Sensible Heat}}{\text{Sensible Heat} + \text{Latent Heat}} = \frac{Q_s}{Q_s + Q_l} \end{aligned}$$

(5.2)

Sensible heat includes internal heat generated in the air-conditioned space and external heat transferred from outdoors to the space. The internal sensible heat is typically generated by

1. People and working equipment.
2. Electric devices.
3. Lighting, etc.

External sensible heat is typically from

1. Solar heat transmitted through the transparent space windows.
2. Heat transfer through the space walls, doors, and ceilings due to the outside and inside temperature difference.
3. Heat brought in with air ventilation, etc.

In general, latent heat comes from inside the space. The heat is typically from

1. People.
2. Cooking appliances, etc.

When the definition of the SHF is applied to room sensible heat, effective sensible heat, and grand sensible heat, the room sensible heat factor (RSHF), the effective sensible heat factor (ESHF), and the grand sensible heat factor (GSHF) are developed. These factors are described in the following sections.

Example 5.1

A theater is measured to have a total heat gain of 3,000 kJ from a heat energy audit. If 600 kJ of the total heat gain is known as latent heat, determine the sensible heat factor of the heat gain in the theater.

SOLUTION

Knowing the total heat and latent heat, the sensible heat is,

$$\text{Sensible heat} = \text{Total heat gain} - \text{Latent heat}$$
$$= 3,000 \text{ kJ} - 600 \text{ kJ} = 2,400 \text{ kJ}$$

Using Equation (5.2), the sensible heat factor is determined to be

$$\text{SHR} = \text{SHF} = \frac{Q_s}{Q_s + Q_l} = \frac{2,400 \text{ kJ}}{3,000 \text{ kJ}} = \mathbf{0.8}$$

5.3.3 ROOM SENSIBLE HEAT FACTOR

The RSHF, also called room sensible heat ratio (RSHR), is defined as the ratio of room sensible heat to total room heat, which is a summation of room sensible heat and room latent heat (RLH). The ratio is expressed as:

$$\text{RSHF} = \frac{\text{Room Sensible Heat}}{\text{Total Room Heat}}$$
$$= \frac{\text{Room Sensible Heat}}{\text{Room Sensible Heat} + \text{Room Latent Heat}} = \frac{Q_{rs}}{Q_{rs} + Q_{rl}} \tag{5.3}$$

The RSHF line represented on the psychrometric chart is a straight line to run parallel with an RSHF base line. The RSHF base line can be drawn by either method 1 or method 2 shown below:

- *Method 1* is to use the half-circle dial, also called protractor, on the top-left side of the chart. The line is drawn from the center of the protractor to the calculated RSHF.
- *Method 2* is to use the axis on the top-right side of the chart. The line is drawn from the alignment point to the calculated RSHF.

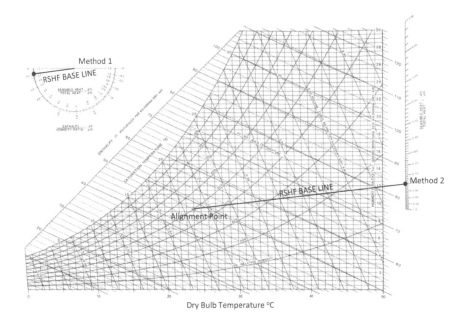

Dry Bulb Temperature °C

FIGURE 5.4 The RSHF base lines.

Figure 5.4 shows the RSHF base lines generated by the two methods. The RSHF base lines are plotted using the SHF 0.8 resulted from *Example 5.1*. The RSHF line starts from the room design condition and ends on the saturation line. In application, the room sensible heat and latent heat should be offset by the conditioned air supplied to the room. Therefore, the RSHF line physically represents a process where the conditioned air is mixed with the room air to reach the room design condition. Figure 5.5 shows the RSHF line drawn by using method 1 represented in Figure 5.4. The room design condition is at 25°C T_{db} and 50% relative humidity (RH).

5.3.4 EFFECTIVE SENSIBLE HEAT FACTOR

The ESHF, also called effective sensible heat ratio (ESHR), is defined as the ratio of effective room sensible heat to total effective sensible heat, which is a summation of effective room sensible heat (ERSH) and effective room latent heat (ERLH). The ratio is expressed as:

$$
\begin{aligned}
\mathrm{ESHF} &= \frac{\text{Effective Room Sensible Heat}}{\text{Effective Total Room Heat}} \\[2mm]
&= \frac{\text{Effective Room Sensible Heat}}{\text{Effective Room Sensible Heat} + \text{Effective Room Latent Heat}} \\[2mm]
&= \frac{\mathrm{ERSH}}{\mathrm{ERSH} + \mathrm{ERLH}} = \frac{Q_{ers}}{Q_{ers} + Q_{erl}}
\end{aligned} \tag{5.4}
$$

The effective room sensible heat consists of room sensible heat Q_{rs} and the portion of outdoor air sensible heat bypassing the coil. The ERLH consists of RLH Q_{rl} and the

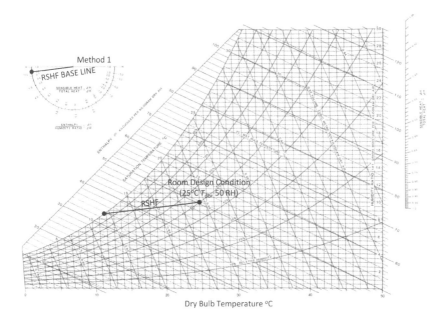

Dry Bulb Temperature °C

FIGURE 5.5 The RSHF line.

portion of outdoor air latent heat bypassing the coil. The bypassed outdoor air load is imposed in a room in the same manner as the infiltration load since the bypassed outdoor air load is supplied to the room through the air distribution system. The ESHF line represented on the psychrometric chart is a straight line generated by using the same methods described for RSHF line. The ESHF line is drawn from the room design condition to the saturation line. Figure 5.6 shows the ESHF base lines and the ESHF line on the psychrometric chart. The ESHF line on the chart is plotted at the ESHF 0.7 and the room design condition at 25°C T_{db} and 50% RH.

The ESHF line may also be drawn on the psychrometric chart without initially using the ESHF base line. In that way, the ESHF line is plotted directly from the room design condition to the saturation line where the GEHF line ends. The GEHF line is described in Section 5.3.6.

5.3.5 Apparatus Dew Point

Apparatus dew point (ADP) is a point used to decide the effective surface temperature of the cooling coil required to accomplish the process of cooling with dehumidifying. This is the temperature that the air would be cooled to at the saturated state, i.e., RH is 100% if the BF is 0. On the psychrometric chart, the ADP is found at the intercept point of the ESHF line and the saturation line. Once the ADP is located, the temperature of the ADP T_{apd} can be determined by plotting a line from ADP vertically to the dry-bulb temperature axis. Then, the volume flow rate $\dot{V}_{da}$ of the dehumidified air passing through the coil can be calculated by using the equation,

$$\dot{V}_{da} = \frac{(\mathrm{ERSH})\, v_{rm}}{c_p \left(1 - \mathrm{BF}\right)\left(T_{rm} - T_{adp}\right)} \tag{5.5}$$

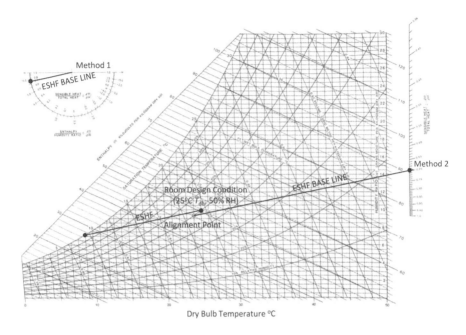

FIGURE 5.6 The ESHF base lines and line.

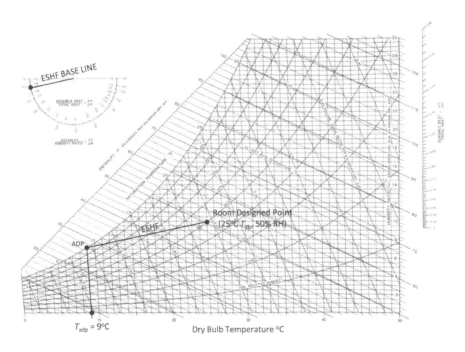

FIGURE 5.7 The ADP and T_{adp}.

where the units of $\dot{V}_{da}$ and ERSH are m³/hr and kJ/hr, respectively. Figure 5.7 shows the ADP on the psychrometric chart. The ADP on the chart is obtained by plotting the ESHF line with the effective sensible heat factor 0.7 and the room design condition at 25°C T_{db} and 50% RH.

Example 5.2

A cooling coil with a 0.05 BF is used in a room air conditioner to keep the room condition at 30°C T_{db} and 60% RH in 1 atm. The air entering the coil is at a rate of 50 m³/min. If knowing the room has an ERSH of 50,100 kJ/hr and ERLH of 15,900 kJ/hr, determine (a) the ESHF, (b) the apparatus dew point temperature T_{adp}, and (c) the volume flow rate of the dehumidified air.

SOLUTION

The state of the air in the room is specified by the given temperature and RH. Referring Appendix A.4 Psychrometric Chart at a Pressure of 1 atm (101.325 kPa), the specific volume is at the state is found,

$$v = 0.8815 \ \frac{m^3}{kg \ dry \ air}$$

Referring Appendix A.3 Specific Heat of Air, the specific heat at the room condition is

$$c_p = 1.006 \ \frac{kJ}{kg.°C}$$

(a) Using Equation (5.4), the ESHF of the room is determined to be

$$ESHF = \frac{ERSH}{ERSH + ERLH} - \frac{50,100 \ \frac{kJ}{hr}}{50,100 \ \frac{kJ}{hr} + 15,900 \ \frac{kJ}{hr}} = 0.76$$

(b) Drawing an ESHF base line on the protractor and plotting a parallel line from the point of the room design condition to intercept with the saturation line, the ADT is found at the cross point. The corresponding T_{apd} is determined on the dry air temperature axis as

$$T_{apd} = \mathbf{19.6°C}$$

(c) Using Equation (5.5), the volume flow rate of the dehumidified air is

$$\dot{V}_{da} = \frac{(ERSH)v_{rm}}{c_p(1 - BF)(T_{rm} - T_{adp})}$$

$$= \frac{\left(50,100 \ \frac{kJ}{hr}\right)\left(0.8815 \ \frac{m^3}{kg \ dry \ air}\right)}{1.006 \ \frac{kJ}{kg.°C}(1 - 0.05)(30 - 19.6)°C} = \mathbf{4,443.3} \ \frac{m^3}{hr}$$

The air properties and process line of Example 5.2 are illustrated in the following psychrometric chart.

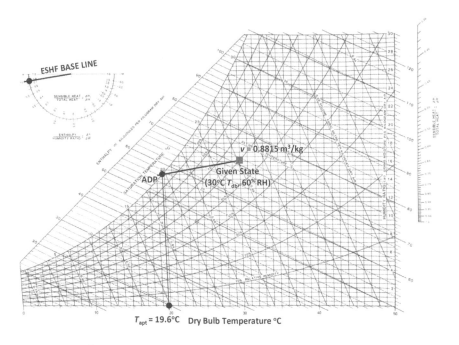

FIGURE Example 5.2

5.3.6 GRAND SENSIBLE HEAT FACTOR

The GSHF, also called grand sensible heat ratio (GSHR), is defined as the ratio of total sensible heat to grand total heat, which is a summation of total sensible heat and total latent heat. The ratio is expressed as:

$$GSHF = \frac{\text{Total Sensible Heat}}{\text{Grand Total Heat}}$$

$$= \frac{\text{Total Sensible Heat}}{\text{Total Sensible Heat} + \text{Total Latent Heat}} = \frac{Q_{ts}}{Q_{ts} + Q_{tl}} \tag{5.6}$$

The grand total heat is the load that the conditioning apparatus must handle, including the outdoor air heat load. Therefore, the state of the air entering the apparatus is a mixture condition of outdoor air and return room air, which physically is the condition of air entering the coil. The condition of the air leaving the apparatus must be located on the GSHF line. The GSHF line represented on the psychrometric chart is a straight line, which can be drawn by the same procedure described for RSHF line. The GSHF line starts from the mixture air condition point and ends at the intercept with the saturation line. The cross point is also the ADP and the end of the ESHF on the saturated line.

Figure 5.8 shows a GSHF line on the psychrometric chart drawn parallel with the GSHF base line on the protractor. The GSHF line on the chart is plotted by using the GSHF of 0.6, the room design condition at 25°C T_{db} and 50% RH, the outdoor air

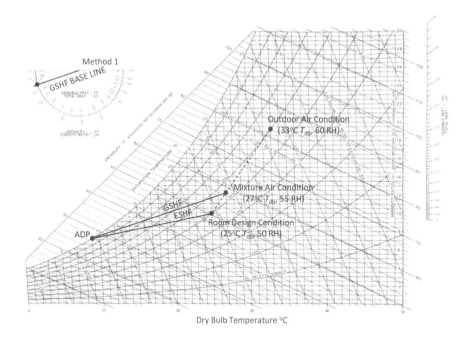

FIGURE 5.8   The GSHF line.

condition at 33°C T_{db} and 60% RH. The air-flow rates from room return and outdoors are 60 m³/min and 20 m³/min, respectively.

5.3.7   EFFECTIVE SURFACE TEMPERATURE

Coil surface temperature varies as air passes through the coil. In coil analysis and calculation, an average constant temperature on the coil surface is typically used. This approximate temperature can be determined by the heat transfer balance of the fluid flowing inside the coil and the air passing outside the coil. The temperature is called effective surface temperature T_{es}. For the air conditioning involving cooling with dehumidifying, the effective surface temperature is at the point where the GSHF line crosses the saturation line on the psychometric chart. Therefore, the effective surface temperature and the dew-point temperature of a coil are the same for the cooling with dehumidifying process, i.e., $T_{dpt} = T_{es}$. For the simple sensible cooling, the effective surface temperature may not be the same as the dew-point temperature.

5.3.8   FACTOR LINES ON THE PSYCHROMETRIC CHART

There are psychrometric relations among the BF, RSHF, ESHF, APT, and GSHF of the coil in air conditioning. The relations can be graphically represented on the psychrometric chart to help analyze the process of cooling with dehumidifying and determine the coil temperatures and load of the air passing through the coil.

In the design specification of air conditioning, the room design condition is given and the outdoor air condition is known. The GSHF, ESHF, and RSHF of the air conditioning are determined by the room energy balance calculation. On the psychrometric chart, the lines of RSHF and ESHF can be specified. Once the supply air flowrate and the condition of air entering the cooling coil are decided, the condition of air leaving the coil can be located on the chart, which is the cross point of the GSHF and RSHF lines. Then, the coil load can be determined accordingly.

Figure 5.9 shows the relations among the factor lines on the psychrometric chart. The outdoor air condition, mixture air condition, and room design condition used in the chart are the same as those in Figure 5.8.

Per definition, the formula of BF is expressed as:

$$BF = \frac{T_{ldb} - T_{adp}}{T_{edb} - T_{adp}} \tag{5.7a}$$

or

$$BF = \frac{\omega_{la} - \omega_{adp}}{\omega_{ea} - \omega_{adp}} \tag{5.7b}$$

or

$$BF = \frac{h_{la} - h_{adp}}{h_{ea} - h_{adp}} \tag{5.7c}$$

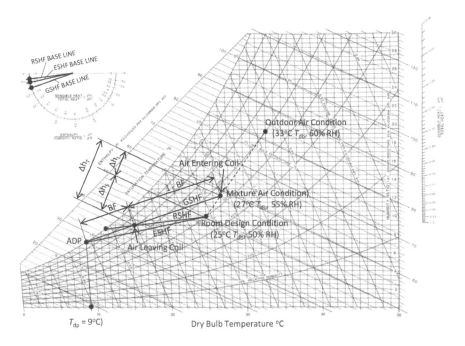

FIGURE 5.9   Factor lines and relations.

where subscripts of l and e represent the states of air leaving and entering the coil. The cooling load of the coil in the process of cooling with humidification can be estimated on the psychrometric chart as shown,

$$\text{Total load} = \text{Sensible load} + \text{Latent load}$$

i.e.

$$\Delta h_T = h_S + h_L$$

in the expression of the specific enthalpies.

5.4 GRAPHIC SOLUTION BY USING THE PSYCHROMETRIC CHART

5.4.1 HEATING PROCESS

As described in Section 5.2, the heating process is a horizontal line on the psychrometric chart that extends to the right from the state of air entering the coil to the state of air leaving the coil. In the heating process, the ESHF and GSHF are 1.0 since no latent heat is involved in the process. The characteristic of the heating process is $\Delta\omega = \omega_2 - \omega_1 = 0$, but the RH changes, $\Delta\phi = \phi_2 - \phi_1 \neq 0$. Figure 5.10 shows a typical

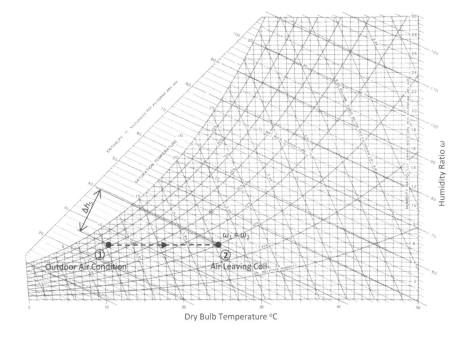

FIGURE 5.10 A heating process of all outdoor air.

heating process of the outdoor air. The heat transfer in the process is calculated from
the energy balance equations, Equations (2.23) and (2.25) (see Chapter 2),

$$\dot{E}_{in} = \dot{E}_{out}$$

$$\dot{Q}_{in} + \dot{m}_{in} h_1 = \dot{m}_{out} h_{12}$$

Since $\dot{Q}_{in} = \dot{Q}$ and mass is conservative

$$\dot{m}_{in} = \dot{m}_{out} = \dot{m}$$

the rate of heat transfer in the heating process is

$$\dot{Q} = \dot{m}(h_2 - h_1)$$

which is the same equation as Equation (2.3b) and the coil load is Δh_S described in
Equation (2.2) (see Chapter 2).

Example 5.3

Outdoor air at a rate of 25 m³/min passes through a heating coil. The air enters the
coil at 20°C DBT and 15°C T_{wb} and leaves the coil at 33°C T_{db}. If the air is a steady
flow at a constant pressure of 100 kPa, determine (a) the enthalpy change (kJ/kg),
(b) the rate of heat transfer (kW), and (c) the change of the absolute humidity and
RH in the process of air passing through the coil.

SOLUTION

The states of the air entering and leaving the coil are specified by the given
temperatures. Referring Appendix A.4 Psychrometric Chart at a Pressure of
1 atm (101.325 kPa), the following properties of the entering and leaving air
are found,

$$h_1 = 42.0 \ \frac{kJ}{kg \ dry \ air}$$

$$v_1 = 0.8421 \ \frac{m^3}{min}$$

$$\phi_1 = 0.5911$$

$$\omega_1 = \omega_2 = 0.0086 \ \frac{kg \ H_2O}{kg \ dry \ air}$$

and

$$h_2 = 55.1 \frac{kJ}{kg \text{ dry air}}$$

$$\phi_2 = 0.2751$$

(a) The enthalpy change of the air is

$$\Delta h = (h_2 - h_1) = (55.1 - 42.0)\frac{kJ}{kg \text{ dry air}} = \mathbf{13.1} \ \frac{\mathbf{kJ}}{\mathbf{kg \ dry \ air}}$$

(b) The mass flow rate is

$$\dot{m} = \frac{\dot{V_1}}{v_1} = \frac{25 \dfrac{m^3}{min}}{0.8421 \dfrac{m^3}{kg}} = 29.69 \ \frac{kg}{min}$$

and the rate of heat transfer in the process of the air passing through the coil becomes

$$\dot{Q} = \dot{m}(h_2 - h_1) = \left(29.69 \ \frac{kg}{min}\right)\left(13.1 \ \frac{kJ}{kg \text{ dry air}}\right)$$

$$= 388.94 \ \frac{kJ}{min} = \frac{388.94 \dfrac{kJ}{min}}{60 \dfrac{s}{min}} = 6.4823 \ \frac{kJ}{s} = \mathbf{6.4823 \ kW}$$

(c) The process is a sensible heating process, the absolute humidity is a constant in the process,

$$\omega_1 = \omega_2 = 0.0086 \ kg \ \frac{kg \ H_2O}{kg \ dry \ air}$$

$$\Delta\omega = \omega_2 - \omega_1 = \mathbf{0}$$

The RH change is

$$\Delta\phi = \phi_2 - \phi_1 = 0.2751 - 0.5911 = -0.316 = \mathbf{-31.6\%}$$

The RH reduces 31.6% in the process.
The air properties and process line of Example 5.3 are illustrated in the following psychrometric chart.

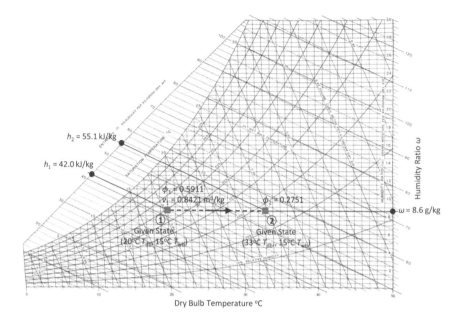

FIGURE Example 5.3

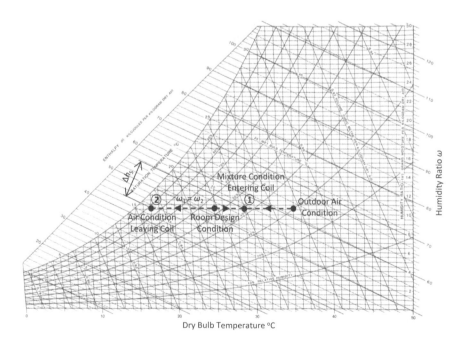

FIGURE 5.11 The cooling process of the mixture air with outdoor and room return air.

5.4.2 COOLING PROCESS

As described in Section 5.2, the cooling process is a horizontal line on the psychro-metric chart, which extends to the left from the state of air entering the coil to the state of air leaving the coil. In the cooling process, the ESHF and GSHF are 1.0 since no latent heat is involved in the process. Similar to the heating process, the charac-teristic of the cooling process is $\Delta\omega = \omega_2 - \omega_2 = 0$, but the RH changes, $\Delta\phi = \phi_2 - \phi_1 \neq 0$ as shown in Figure 5.10. The coil load is Δh_S. Figure 5.11 shows a typical cooling process of the mixture air with the outdoor and room return air.

Example 5.4

Air passing through a cooling coil in a department store is sent to a storage room. The process is sensible cooling. The temperature in the room is kept at 22°C T_{db}. The air supply is 100% from ventilation at a rate of 370 m³/min from the outdoor at 40°C T_{db} and 21°C T_{wb}. If knowing that the BF of the coil is 0.05 and the room has 210,000 kJ/hr of the sensible heat, determine (a) the effective sensible heat factor (ESHF), (b) the coil dew point (T_{apt}), (c) the dehumidified air quantity $\dot{V}_{da}$, (d) the effective surface temperature (T_{es}), (e) the supply air temperature (T_{sa}), and (f) the cooling coil load (kW) of the process.

SOLUTION

The state of the outside air is specified by the given temperatures. The state of the air in the storage room is specified by the given temperature and specific humidity. Referring to Appendix A.4 Psychrometric Chart at a Pressure of 1 atm (101.325 kPa), the following properties of the air entering the coil and the room air are found,

$$h_1 = 60.4 \ \frac{kJ}{kg \ dry \ air}$$

$$V_1 = 0.8886 \ \frac{m^3}{kg}$$

and

$$h_2 = 42.1 \frac{kJ}{kg \ dry \ air}$$

$$V_2 = 0.8452 \ \frac{m^3}{kg}$$

The mass flow rate in the process, therefore, is

$$\dot{m} = \frac{\dot{V}_1}{V_1} = \frac{370 \ \frac{m^3}{min}}{0.8886 \ \frac{m^3}{kg}} = 416.39 \ \frac{kg}{min}$$

The outdoor sensible heat load is

$$\dot{Q} = \dot{m}(h_1 - h_2) = \left(416.39 \ \frac{kg}{min}\right)(60.4 - 42.1) \ \frac{kJ}{kg \ dry \ air}$$

$$= 7,619.94 \ \frac{kJ}{min} = \left(7,619.94 \ \frac{kJ}{min}\right)\left(60 \ \frac{min}{hr}\right) = 457,196.4 \ kJ/hr$$

(a) There is no latent heat involved in the process. From Equation 5.4, the ESHF is

$$ESHF = \frac{210,000\ \dfrac{kJ}{hr} + 0.05\left(457,196.4\ \dfrac{kJ}{hr}\right)}{210,000\ \dfrac{kJ}{hr} + 0.05\left(457,196.4\ \dfrac{kJ}{hr}\right)} = 1.0$$

Therefore, the ESHF line is a horizontal line represented on the psychrometric chart. The lines of ESHF and RSHF are identical.

(b) Plotting the ESHF from the room condition toward the saturation line on the chart, the T_{adp} is read on the dry-bulb axis,

$$T_{adp} = 10.2°C$$

The enthalpy at the T_{adp} is found as

$$h_{adp} = 29.6\ kJ/kg$$

(c) Using Equation (5.5), the dehumidified air quantity $\dot{V}_{da}$ can be determined,

$$\dot{V}_{da} = \frac{(ERSH)v_{rm}}{c_p(1-BF)(T_{rm}-T_{adp})}$$

$$= \frac{\left[210,000\ \dfrac{kJ}{hr} + 0.05\left(457,196.4\ \dfrac{kJ}{hr}\right)\right]0.8452\ \dfrac{m^3}{kg}}{\left(1.006\ \dfrac{kJ}{kg.K}\right)(1-0.05)(22-10.2)°C}$$

$$= 17,452.2\ \frac{m^3}{hr} = \frac{17,452.2\ \dfrac{m^3}{hr}}{60\ \dfrac{min}{hr}} = 290.87\ m^3/min$$

(d) Since $\dot{V}_{da}$ is less than the outdoor ventilation 370 m³/min, the effective surface temperature T_{es} is determined based on the outdoor ventilation,

$$T_{es} = T_2 - \frac{(ERSH)v_{rm}}{c_p(1-BF)\dot{V}_{vent}}$$

$$= 22°C - \frac{\left[210,000\ \dfrac{kJ}{hr} + 0.05\left(457,196.4\ \dfrac{kJ}{hr}\right)\right]0.8452\ \dfrac{m^3}{kg}}{\left(1.006\ \dfrac{kJ}{kg.K}\right)(1-0.05)\left(370\ \dfrac{m^3}{min}\right)\left(60\ \dfrac{min}{hr}\right)}$$

$$= 22°C - 9.23°C = 12.8°C$$

The effective surface temperature T_{es} falls on the ESHF line.

(e) Using Equation (5.7a),

$$BF = \frac{T_{ldb} - T_{adp}}{T_{edb} - T_{adp}}$$

the air temperature T_{ldb} leaving the coil is determined to be

$$T_{ldb} = BF\left(T_{edb} - T_{adp}\right) + T_{es}$$

T_{adp} is replaced by T_{es} in the simple cooling application. T_{sa} is the same as the T_{ldb}, then

$$T_{sa} = 0.05(40 - 12.8)°C + 12.8°C = \mathbf{14.2°C}$$

(f) The cooling load is determined to be

$$\dot{Q}_{cooling} = \dot{m}(h_1 - h_3) = \left(416.39 \ \frac{kg}{min}\right)(60.4 - 34.4)\frac{kJ}{kg \ dry \ air}$$

$$= 10{,}826.1 \ \frac{kJ}{min} = \frac{10{,}826.1 \ \dfrac{kJ}{min}}{60 \ \dfrac{s}{min}} = 180.4 \ \frac{kJ}{s} = \mathbf{180.4 \ kW}$$

The air properties and process lines of Example 5.4 are illustrated in the following psychrometric chart.

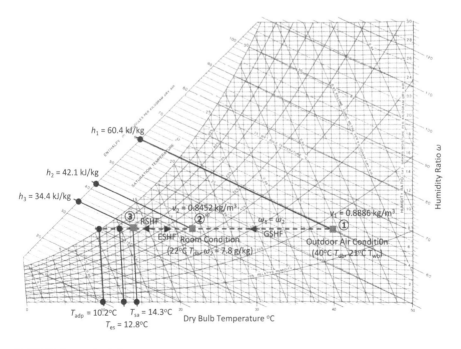

FIGURE Example 5.4

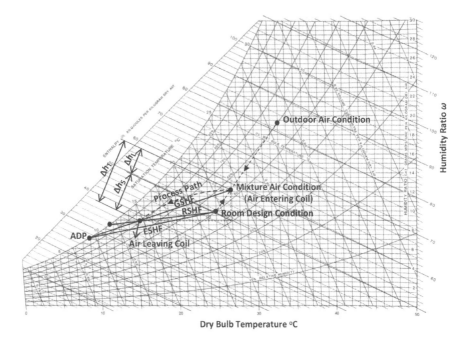

FIGURE 5.12   The cooling with dehumidifying process.

5.4.3 COOLING WITH THE DEHUMIDIFYING PROCESS

In the calculation and analysis of air conditioning, one important task is to determine the coil load. The load can be conveniently determined by using the chart once the positions of the air entering and exiting the coil on the chart are identified,

$$\dot{Q}_{\text{cool}} = \dot{m}(h_1 - h_2) = \dot{m}h_{\text{T}} = \dot{m}(\Delta h_{\text{s}} + \Delta h_{\text{l}})$$

where h_1 and h_2 are the enthalpies of air entering and leaving the coil. The load of chilled water or refrigerant inside the coils can be calculated by

$$\text{Tonnage} = \frac{\dot{Q}_{\text{cool}}\ (\text{kW})}{3.519\ (\text{kW/ton})} \tag{5.8}$$

Figure 5.12 shows the process and factor lines of cooling with dehumidifying on the psychrometric chart.

Example 5.5

A chilled-water cooling coil in a bank is used to cool air with dehumidifying. The air passing through the coil is sent to an air-conditioned space at 23°C T_{db} and 50% RH. The outdoor air condition around the space is 37°C T_{db} and 24°C T_{wb}.

The ventilation rate from outside is 60 m³/min. If the coil has a BF of 0.15 and the space has 210,000 kJ/hr RSH and 43,000 kJ/hr of the RLH based on the heat balance audit, determine (a) the outdoor air heat load, (b) the RSHF, (c) the grand total heat, (d) the ESHF, (e) the T_{adp}, (f) the dehumidified air quantity, (g) the air flow rate entering the coil, (h) T_{edb} and T_{ewb} of the air entering the coil, (i) T_{ldb} and T_{lwb} of the air leaving the coil, and (j) the chilled water tonnage of the process.

SOLUTION

The states of the outdoor air and the space air are specified by the given temperatures and RH, respectively. Referring to Appendix A.4 Psychrometric Chart at a Pressure of 1 atm (101.325 kPa), the following properties of the outdoor air and the space air are found,

$$h_1 = 71.9 \ \frac{kJ}{kg}$$

$$v_1 = 0.8882 \ \frac{m^3}{kg \ dry \ air}$$

and

$$h_2 = 45.2 \ \frac{kJ}{kg}$$

$$v_2 = 0.8512 \ \frac{m^3}{kg \ dry \ air}$$

The mass flow rates of the outdoor air is determined,

$$\dot{m}_{oa} - \frac{\dot{V}_1}{v_1} - \frac{60 \ \frac{m^3}{min}}{0.8882 \ \frac{m^3}{kg}} = 67 \ 55 \ \frac{kg}{min}$$

(a) Reading from the chart, the latent heat of the outdoor air is

$$\dot{Q}_l = \dot{m}_1(h_1 - h) = \left(67.55 \ \frac{kg}{min} \right)(71.9 - 59.2) \ \frac{kJ}{kg}$$

$$= 857.9 \frac{kJ}{min} = 51,474 \ \frac{kJ}{hr}$$

And the sensible heat of the outdoor air is

$$\dot{Q}_s = \dot{m}_1(h - h_2) = \left(67.55 \ \frac{kg}{min} \right)(59.2 - 45.2) \ \frac{kJ}{kg}$$

$$= 945.7 \ \frac{kJ}{min} = 56,742 \ \frac{kJ}{hr}$$

The total outdoor air heat load is determined,

$$\dot{Q}_{oa} = \dot{Q}_l + \dot{Q}_s = 51,474 \, \frac{kJ}{hr} + 56,742 \, \frac{kJ}{hr}$$

$$= 108,216 \, \frac{kJ}{hr}$$

(b) Using Equation (5.3), the RSHF is

$$RSHF = \frac{\text{Room Sensible Heat}}{\text{Room Sensible Heat} + \text{Room Latent Heat}}$$

$$= \frac{210,000 \, \frac{kJ}{hr}}{210,000 \, \frac{kJ}{hr} + 43,000 \, \frac{kJ}{hr}} = 0.83$$

(c) The total sensible heat is the RSH plus the outdoor air sensible heat $\dot{Q}_s$

$$\dot{Q}_{total,s} = RSH + \dot{Q}_s$$

$$= 210,000 \, \frac{kJ}{hr} + 56,742 \, \frac{kJ}{hr} = 266,742 \, \frac{kJ}{hr}$$

And the total latent heat is the RLH plus the outdoor air latent heat $\dot{Q}_l$,

$$\dot{Q}_{total,l} = RLH + \dot{Q}_l$$

$$= 43,000 \, \frac{kJ}{hr} + 51,474 \, \frac{kJ}{hr} = 94,474 \, \frac{kJ}{hr}$$

The GTH is determined to be

$$GTH = \dot{Q}_{total,s} + \dot{Q}_{total,l}$$

$$GTH = 266,742 \, \frac{kJ}{hr} + 94,474 \, \frac{kJ}{hr} = 361,216 \, \frac{kJ}{hr}$$

(d) Using Equation (5.4), the ESHF is determined,

$$ESHF = \frac{\text{Effective Room Sensible Heat}}{\text{Effective Room Sensible Heat} + \text{Effective Room Latent Heat}}$$

$$= \frac{210,000 \, \frac{kJ}{hr} + 0.15 \left(56,742 \, \frac{kJ}{hr} \right)}{210,000 \, \frac{kJ}{hr} + 0.15 \left(56,742 \, \frac{kJ}{hr} \right) + 43,000 \, \frac{kJ}{hr} + 0.15 \left(51,474 \, \frac{kJ}{hr} \right)}$$

$$= 0.81$$

(e) The T_{adp} is obtained to find ADP by drawing the ESHF line from the room design condition toward the saturation line on the psychrometric chart. The T_{adp} is read on the dry-bulb temperature axis,

$$T_{adp} = \textbf{8.7°C}$$

(f) Using Equation (5.5), the dehumidified air quantity is determined,

$$\dot{V}_{da} = \frac{(ERSH)\,v_{rm}}{c_p\,(1 - BF)\,(T_{rm} - T_{adp})}$$

$$= \frac{\left[210{,}000\,\dfrac{kJ}{hr} + 0.15\left(56{,}742\,\dfrac{kJ}{hr}\right)\right]0.8512\,\dfrac{m^3}{kg}}{1.006\,\dfrac{kJ}{(kg.K)}(1 - 0.15)(23 - 8.7)°C}$$

$$= 15.210.8\,\frac{m^3}{hr} = \textbf{253.5}\,\frac{\textbf{m}^3}{\textbf{min}}$$

(g) Using the BE and $\dot{V}_{da}$, the air flow rate entering the coil $\dot{V}_{sa}$ is obtained by

$$\dot{V}_{sa} = \frac{\dot{V}_{da}}{1 - BF} = \frac{253.5\,\dfrac{m^3}{min}}{1 - 0.15} = 298.2\,\frac{m^3}{min}$$

Knowing 60 m³/min is the air flow rate from the outdoor, 238.2 m³/min should be the air flow rate from the room return. The air mass flow rate from the room return is,

$$\dot{m}_{ra} = \frac{238.2\,\dfrac{m^3}{min}}{0.8512\,\dfrac{m^3}{kg}} = 279.8\,\frac{kg}{min}$$

The total mass flow rate of the air entering the coils, which is the mixed air from the outdoor and room air return is determined to be

$$\dot{m}_{ma} = \dot{m}_{oa} + \dot{m}_{ra} = 67.55\,\frac{kg}{min} + 279.8\,\frac{kg}{min} = 347.35\,\frac{kg}{min}$$

(h) From Equation (3.15) (see Chapter 3), the T_{edb} at the coil is determined to be

$$T_{edb} = \frac{\dot{m}_{OA}\,T_1 + \dot{m}_{RA}\,T_2}{\dot{m}_{ma}}$$

$$= \frac{\left(67.55\,\dfrac{kg}{min}\right)(37°C) + \left(279.8\,\dfrac{kg}{min}\right)(23°C)}{347.35\,\dfrac{kg}{min}}$$

$$= \textbf{25.7°C}\ T_{db}$$

Reading T_{ewb}, which is on the line between the states of the outdoor air and the room design condition on the psychrometric chart, T_{ewb} is

$$T_{ewb} = \textbf{17.8°C}$$

The GSHF line is obtained by drawing the line from the mixing condition of the air entering the coil to the T_{adp}.

(i) The T_{ldb} is determined by calculation,

$$T_{ldb} = T_{adp} + BF\left(T_{edb} - T_{adp}\right)$$

$$T_{ldb} = 8.7 + 0.15(25.7 - 8.7)°C = \textbf{11.25°C db}$$

Alternatively, the T_{ldb} can be found at the intercept point between the GSHF line and the ESHF line. The T_{lwb} is read at the intercept point as

$$T_{lwb} = \textbf{10.5°C}$$

(j) The tonnage of chilled water in the process is determined from Equation (5.8). In the equation

$$\dot{Q}_{cool} = GTH = 361,216 \ \frac{kJ}{hr}$$

Alternatively, $\dot{Q}_{cool}$ can be determined by using Equation (2.3b) (see Chapter 2), i.e.,

$$\dot{Q}_{cool} = \dot{m}_{ma}\left(h_{ma} - h_{la}\right) = 347.35 \frac{kg}{min}(49.5 - 31.0)\frac{kJ}{kg}$$

$$= 6,426.0 \ \frac{kJ}{min} = 385,558.5 \ \frac{kJ}{hr}$$

The tonnage of chilled water using the GTH is

$$Tonnnage = \frac{\dot{Q}_{cool} \ kW}{3.517 \ \dfrac{kW}{ton}} = \frac{\left(361,216 \ \dfrac{kJ}{hr}\right)\Big/\left(3,600 \ \dfrac{s}{hr}\right)}{3.517 \left(\dfrac{kW}{ton}\right)}$$

$$= 28.53 \ tons = \textbf{29 tons}$$

Or from the above calculation, the tonnage of chilled water is

$$Tonnage = \frac{\dot{Q}_{cool} \ kW}{3.517 \ \dfrac{kW}{ton}} = \frac{\left(385,558.5 \ \dfrac{kJ}{hr}\right)\Big/\left(3,600 \ \dfrac{s}{hr}\right)}{3.517 \left(\dfrac{kW}{ton}\right)}$$

$$= 30.45 \ tons = \textbf{31 tons}$$

The error in percentage of the tonnage difference by using the GTH and above calculation is

$$\Delta_{Tonnage}(\%) = \frac{(31-29)\,\text{tons}}{29\ \text{tons}} = 0.068 = 6.8\%$$

The air properties and process lines of Example 5.5 are illustrated in the following psychrometric chart.

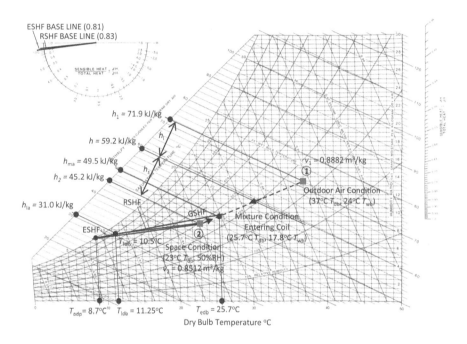

FIGURE Example 5.5

5.5   AIR CHANGE RATE

One of the purposes of applying the heat gain method in air conditioning design is to determine the rate of supply air, also called the airflow load. The airflow load provides a base for selecting the air conditioning equipment and designing the fluid flow system. The heat gain method needs to consider both the external heat load and internal heat load of RSH and RLH. Sometimes, however, air conditioning design is concerned with providing the air-conditioned space good air quality, complying with HAVC standards or regulations. Then, the method of air change rate (ACH) is more suitable to be used alternatively.

The method of ACH is to measure the number of times of the air within an air-conditioned space replaced every hour. After the ACH of the air-conditioned space is decided, the airflow load $\dot{G}$ (m³/hr) entering the air-conditioned space is identified, i.e.,

$$\dot{G} = \text{ACH} \times V \tag{5.9}$$

where ACH and V are the air change rate and volume of the air-conditioned space in units of times/hr and m³/time, respectively. The frequency of air change out Mn in minutes of the space, therefore, is determined to be

$$M_n = \frac{60 \text{ min}}{N \text{ times}} \quad (5.10)$$

For example, the rate of supply air to a room that is a volume of 1,000 m³ with 4 ACH can be decided by Equation (5.9),

$$\dot{G} = \left(4 \frac{\text{times}}{\text{hr}}\right)\left(1,000 \frac{\text{m}^3}{\text{time}}\right) = 4,000 \frac{\text{m}^3}{\text{hr}} = 66.67 \frac{\text{m}^3}{\text{min}}$$

and using Equation (5.10), the frequency of air change out for the room is

$$M_n = 60 \text{ min}/4 \text{ times} = 15 \frac{\text{min}}{\text{time}}$$

Air change rate is typically recommended based on the space ventilation requirement. The air change rate of a laboratory may not be the same as that of a residential room. The former is in consideration of the dilution of known contaminant and the latter is in consideration of number of occupants in the space. Many regulatory bodies recommend ACH rates for various types of spaces to ensure enough air changes for optimal air quality. Table 5.1 lists the recommended air change rates of various spaces.

TABLE 5.1
Recommended Air Change Rates(ACH)

Building/Room	ACH (Times/Hour)	Building/Room	ACH (Times/Hour)
All spaces in general	min 4	Night clubs	20–30
Assembly halls	4–6	Machine shops	6–12
Attic spaces for cooling	12–15	Malls	6–10
Auditoriums	8–15	Medical centers	8–12
Bakeries	2–30	Medical clinics	8–12
Banks	4–10	Medical offices	8–12
Barber shops	6–10	Mills, paper	15–20
Bars	20–30	Mills, textile general buildings	4
Bathrooms	10–20	Mills, textile dye houses	15–20
Beauty shops	6–10	Municipal buildings	4–10
Boiler rooms	15–20	Museums	12–15

(Continued)

TABLE 5.1 *(Continued)*
Recommended Air Change Rates(ACH)

Building/Room	ACH (Times/Hour)	Building/Room	ACH (Times/Hour)
Bowling alleys	10–15	Offices, public	3
Cafeterias	12–15	Offices, private	4
Churches	8–15	Paint shops	10–15
Classrooms	6–20	Photo dark rooms	10–15
Club rooms	2	Pig houses	6–10
Club houses	20–30	Police stations	4–10
Cocktail lounges	20–30	Post offices	4–10
Computer rooms	15–20	Poultry houses	6–10
Court houses	4–10	Precision manufacturing	10–15
Dance halls	6–9	Pump rooms	5
Dental centers	8–12	Railroad shops	4
Department stores	4–10	Residences	1–2
Dining halls	12–15	Restaurants	8–12
Dining rooms (restaurants)	12–15	Retail	6–10
Dress shops	6–10	School classrooms	4–12
Drug shops	6–10	Shoe shops	6–10
Engine rooms	4–6	Shopping centers	6–10
Factory buildings, ordinary	2–4	Shops, machine	5
Factory buildings, with fumes/moisture	10–15	Shops, paint	15–20
Fire stations	4–10	Shops, woodworking	5
Foundries	15–20	Substation, electric	5–10
Galvanizing plants	20–30	Supermarkets	4–10
Garages repair	20–30	Swimming pools	20–30
Garages storage	4–6	Town halls	4–10
Homes, night cooling	10–18	Taverns	20–30
Hospital rooms	4–6	Theaters	8–15
Jewelry shops	6–0	Transformer rooms	10–30
Kitchens	15–60	Turbine rooms, electric	5–10
Laundries	10–15	Warehouses	2
Libraries, public	4	Waiting rooms, public	4
Lunchrooms	12–15	Warehouses	6–30
Luncheonettes	12–15	Wood-working shops	8

Example 5.6

An air conditioner is designed to supply conditioned air to an industrial build-ing. The outdoor air condition around the building is 35°C T_{db} and 27°C T_{wb}. The condition in the build is kept at 22°C T_{db} and 50% RH. The conditioner is required to have the condition of air leaving the cooling coil at 9.2°C T_{db} and

8.3°C T_{db}. The building rooms and the ACH requested for each room are shown in the table.

No.	Room Name	Dimensions (L × W × H)	ACH
1	Mechanical room	10 × 8 × 5	18
2	Cafeteria	12 × 8 × 5	13
3	Conference room	8 × 8 × 3	10
4	Open employee office	20 × 16 × 5	4
5	Manager office	5 × 5 × 3	4
6	Office 1	3 × 5 × 3	4
7	Office 2	3 × 5 × 3	4
8	Library	8 × 5 × 5	4
9	Men's room	6 × 3 × 3	12
10	Women's room	5 × 3 × 3	12
11	Storage room	5 × 5 × 3	8
12	Workshop	10 × 8 × 5	10
13	Mailroom	5 × 3 × 3	6
14	Hallway	30 × 3 × 5	5
15	Reception desk area	5 × 3 × 3	4
16	Printer room	5 × 5 × 3	15
17	Janitor room	3 × 3 × 3	5
18	IT room	3 × 5 × 3	15

The room air distribution is specified as below.

1. Mechanical room: 50% supply air, 50% transient air from cafeteria, and 100% exhaust air.
2. Cafeteria: 30% return of the supply air and other exhaust air except the transient air to mechanical room.
3. Men's room: 100% transient air from hallway and 100% exhaust air.
4. Women's room: 100% transient air from hallway and 100% exhaust air.
5. Janitor room: 100% transient air from hallway and 100% exhaust air.
6. Workshop: 50% supply air, 50% transient air from open employee office, and 100% exhaust air.

If the building has one HVAC system distributing the conditioned air to all rooms, determine the capacity of (a) supply air (SA), (b) the exhaust air (EA), (c) the return air (RA), (d) the outside air (OA), (e) the coil cooling load (kW) and the chilled water capacity (tons), and (f) sketch the air flow diagram of the air distribution.

SOLUTION

According to the given room dimension and ACH, the air requirement of each room is determined:

$$\dot{V}_{\text{Mechanical Room}} = \left(18\ \frac{\text{times}}{\text{hr}}\right)(10 \times 8 \times 5)\frac{\text{m}^3}{\text{time}} = 7{,}200\ \frac{\text{m}^3}{\text{hr}}$$

$$\dot{V}_{\text{Cafeteria}} = \left(13 \ \frac{\text{times}}{\text{hr}}\right)(12 \times 8 \times 5)\frac{m^3}{\text{time}} = 6,420 \ m^3/h$$

$$\dot{V}_{\text{Conference Room}} = \left(10 \ \frac{\text{times}}{\text{hr}}\right)(8 \times 8 \times 3)\frac{m^3}{\text{time}} = 1,920 \ \frac{m^3}{\text{hr}}$$

$$\dot{V}_{\text{Open Employee Office}} = \left(4 \ \frac{\text{times}}{\text{hr}}\right)(20 \times 16 \times 5)\frac{m^3}{\text{time}} = 6,400 \ \frac{m^3}{\text{hr}}$$

$$\dot{V}_{\text{Manager Office}} = \left(4 \ \frac{\text{times}}{\text{hr}}\right)(5 \times 5 \times 3)\frac{m^3}{\text{time}} = 300 \ \frac{m^3}{\text{hr}}$$

$$\dot{V}_{\text{Office 1}} = \left(4 \ \frac{\text{times}}{\text{hr}}\right)(3 \times 5 \times 3)\frac{m^3}{\text{time}} = 180 \ \frac{m^3}{\text{hr}}$$

$$\dot{V}_{\text{Office 2}} = \left(4 \ \frac{\text{times}}{\text{hr}}\right)(3 \times 5 \times 3)\frac{m^3}{\text{time}} = 180 \ \frac{m^3}{\text{hr}}$$

$$\dot{V}_{\text{Library}} = \left(4 \ \frac{\text{times}}{\text{hr}}\right)(8 \times 5 \times 5)\frac{m^3}{\text{time}} = 800 \ \frac{m^3}{\text{hr}}$$

$$\dot{V}_{\text{Men's Room}} = \left(12 \ \frac{\text{times}}{\text{hr}}\right)(6 \times 3 \times 3)\frac{m^3}{\text{time}} = 648 \ \frac{m^3}{\text{hr}}$$

$$\dot{V}_{\text{Women's Room}} = \left(12 \ \frac{\text{times}}{\text{hr}}\right)(5 \times 3 \times 3)\frac{m^3}{\text{time}} = 540 \ \frac{m^3}{\text{hr}}$$

$$\dot{V}_{\text{Storage Room}} = \left(8 \ \frac{\text{times}}{\text{hr}}\right)(5 \times 5 \times 3)\frac{m^3}{\text{time}} = 600 \ \frac{m^3}{\text{hr}}$$

$$\dot{V}_{\text{Workshop}} = \left(10 \ \frac{\text{times}}{\text{hr}}\right)(10 \times 8 \times 5)\frac{m^3}{\text{time}} = 4,000 \ \frac{m^3}{\text{hr}}$$

$$\dot{V}_{\text{Mailroom}} = \left(6 \ \frac{\text{times}}{\text{hr}}\right)(5 \times 3 \times 3)\frac{m^3}{\text{time}} = 270 \ \frac{m^3}{\text{hr}}$$

$$\dot{V}_{\text{Hallway}} = \left(5 \ \frac{\text{times}}{\text{hr}}\right)(30 \times 3 \times 5)\frac{m^3}{\text{time}} = 2,250 \ \frac{m^3}{\text{hr}}$$

$$\dot{V}_{\text{Reception Desk Area}} = \left(4 \ \frac{\text{times}}{\text{hr}}\right)(5 \times 3 \times 3)\frac{m^3}{\text{time}} = 180 \ \frac{m^3}{\text{hr}}$$

$$\dot{V}_{\text{Printer Room}} = \left(15 \ \frac{\text{times}}{\text{hr}}\right)(5 \times 5 \times 3)\frac{m^3}{\text{time}} = 1,125 \ \frac{m^3}{\text{hr}}$$

$$\dot{V}_{\text{Janitor Room}} = \left(5 \ \frac{\text{times}}{\text{hr}}\right)(3 \times 3 \times 3)\frac{m^3}{\text{time}} = 135 \ \frac{m^3}{\text{hr}}$$

$$\dot{V}_{\text{IT Room}} = \left(15 \ \frac{\text{times}}{\text{hr}}\right)(3 \times 5 \times 3)\frac{m^3}{\text{time}} = 675 \ \frac{m^3}{\text{hr}}$$

(a) Since the mechanical room and workshop have 50% supply air, and the men's room, women's room, and janitor room don't have air supply, the total supply air capacity from the air conditioner is to add the air distribution to all room $\dot{V}_{all}$ except above the room's air capacities, i.e.,

$$\dot{V}_{sa} = \dot{V}_{all} - 50\%\left(\dot{V}_{Mechanical\ Room} + \dot{V}_{Workshop}\right) - \left(\dot{V}_{Men's\ Room} + \dot{V}_{Women's\ Room} + \dot{V}_{Janitor\ Room}\right)$$

$$= 33,823\,\frac{m^3}{hr} - 0.5(7,200 + 4,000)\frac{m^3}{hr} - (648 + 540 + 135)\frac{m^3}{hr}$$

$$= (33,823 - 5,600 - 1,323)\frac{m^3}{hr} = \mathbf{26,900}\,\frac{m^3}{hr}$$

(b) The capacity of exhaust air will be:

$$\dot{V}_{ea} = \dot{V}_{Mechanical\ Room} + \dot{V}_{Workshop} + \left(\dot{V}_{Cafeteria} - 0.3\dot{V}_{Cafeteria} - 0.5\dot{V}_{Mechanical\ Room}\right)$$

$$+ \dot{V}_{Men's\ Room} + \dot{V}_{Women's\ Room} + \dot{V}_{Janet\ Room}$$

$$= \left[7,200 + 4,000 + (6,420 - 0.3 \times 6,420 - 0.5 \times 7,200) + 648 + 540 + 135\right]\frac{m^3}{hr}$$

$$= (11,200 + 894 + 1,323)\frac{m^3}{hr} = \mathbf{13,417}\,\frac{m^3}{hr}$$

(c) Based on the mass flow conservative, capacity of the return air should be equal to the capacity of the supply air minus the capacity of the exhaust air, i.e.,

$$\dot{V}_{ra} = \dot{V}_{sa} - \dot{V}_{ea} = (26,900 - 13,417)\frac{m^3}{hr} = \mathbf{13,483}\,\frac{m^3}{hr}$$

(d) The capacity of the outside air should be equal to the capacity of the exhaust air, i.e.,

$$\dot{V}_{oa} = \dot{V}_{ea} = \mathbf{13,417}\,\frac{m^3}{hr}$$

(e) The coil cooling load is determined by

$$\dot{Q}_{cool} = \dot{m}(h_1 - h_2)$$

In the equation, $\dot{m}$ is the mass flow rate of air entering the coil. Referring to Appendix A.4 Psychrometric Chart at a Pressure of 1 atm (101.325 kPa), the specific volume of outdoor air and return air are, respectively,

$$v_{oa} = 0.90\,\frac{m^3}{kg}$$

$$v_{ra} = 0.8468\,\frac{m^3}{kg}$$

Then, the mass flow rates of outdoor air and return air are determined to be

$$\dot{m}_{oa} = \frac{\dot{V}_{oa}}{v_{oa}} = \frac{13{,}417 \ \dfrac{m^3}{hr}}{0.90 \ \dfrac{m^3}{\text{kg dry air}}} = 14{,}907.8 \ \frac{\text{kg dry air}}{hr}$$

$$\dot{m}_{ra} = \frac{\dot{V}_{ra}}{v_{ra}} = \frac{13{,}483 \ \dfrac{m^3}{hr}}{0.8468 \ \dfrac{m^3}{\text{kg dry air}}} = 15{,}922.3 \ \frac{\text{kg dry air}}{hr}$$

Using Equation (3.15) (see Chapter 3), the temperature of the supply air, i.e., the temperature of air entering coil is,

$$T_1 = \frac{\dot{m}_{oa}T_{oa} + \dot{m}_{ra}T_{ra}}{\dot{m}_{oa} + \dot{m}_{ra}}$$

$$= \frac{\left(14{,}907.8 \ \dfrac{\text{kg dry air}}{hr}\right)35°C + \left(15{,}922.3 \ \dfrac{\text{kg dry air}}{min}\right)22°C}{(14{,}907.8 + 15{,}922.3)\dfrac{\text{kg dry air}}{min}} = 28.3°C$$

The position of the mixture air represented on the psychrometric chart is on the line connecting the conditions of outdoor air and room air. From the chart, enthalpies of air entering and leaving the coil are found, respectively

$$h_1 = 65.2 \ \frac{kJ}{\text{kg dry air}}$$

$$h_2 = 25.5 \ \frac{kJ}{\text{kg dry air}}$$

The coil cooling load, therefore, is

$$\dot{Q}_{cool} = (14{,}907.8 + 15{,}922.3) \ \frac{\text{kg dry air}}{min}(65.2 - 25.5) \ \frac{kJ}{\text{kg dry air}}$$

$$= 1{,}223{,}955.0 \ \frac{kJ}{min} = \frac{1{,}223{,}955.0 \ \dfrac{kJ}{min}}{60 \ \dfrac{s}{min}} = 20{,}399.3 \ \frac{kJ}{s} = \mathbf{20{,}399.3 \ kW}$$

(f) The sketch of the air flow diagram of the air conditioning is shown below.
The air properties and process lines of Example 5.6 are illustrated in the following psychrometric chart.

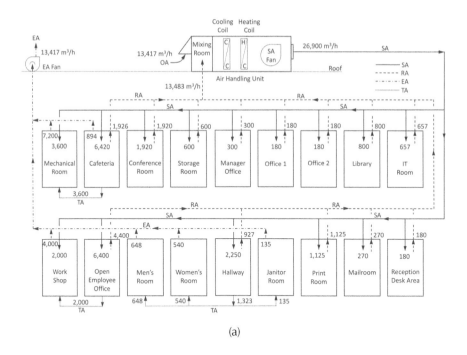

(a)

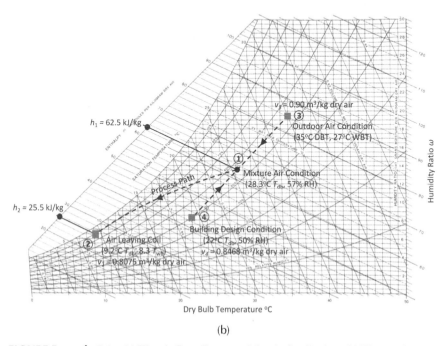

(b)

FIGURE Example 5.6 (a) The air flow diagram of the air distribution. (b) The psychrometric chart.

6 Coil Fluid Systems

6.1 INTRODUCTION

A coil fluid system is constructed to transport fluid to heating or cooling coils. The system has two fluid flows: one is an external flow, which is the air passing through the coil, and another is an internal fluid flow inside the coil, such as hot water, steam, refrigerant, or chilled water. The fluid circulates in a piping system or tubing loop connecting to the coil. Equipment and devices required in the system are varied depending on the fluid and process. The equipment and devices required in a heating system are not the same as those required in a cooling system. Even for the same heating purpose, the equipment and devices required in the steam heating system differ from that in the hot water system. Figure 6.1 shows a schematic of a coil fluid piping system connecting to heating and cooling coils. In the system, steam is from a steam boiler. The refrigerant and chilled water are from an electric chiller and absorption chiller, respectively, for cooling. Power generated from a turbine-generator is used to drive the electric chiller and other electric devices, such as fans and pumps.

6.2 HEATING SYSTEMS

In heating systems, air passing through the heating coil absorbs heat from hot fluid flowing inside the coil so that the air temperature increases. The hot fluid can be hot water or steam generated in a boiler, heat exchanger, or other heat resource. Figure 6.2 shows schematics of the hot water and steam piping systems connecting to the heating coil.

6.3 COOLING SYSTEMS

In cooling systems, air passing through the cooling coil releases heat to the cold fluid flowing inside the coil. So the air temperature gets down. The cold fluid can be refrigerant or chilled water generated in an electric chiller or absorption chiller. Figure 6.3 shows the refrigerant and chilled water piping systems connected to the cooling coil.

6.4 EQUIPMENT AND DEVICES IN SYSTEMS

To perform proper and efficient heating and cooling processes, the installation of various well-functioning equipment and devices is necessary in coil fluid piping systems. The major equipment and devices involved in fluid piping systems, such as air handling units, fan coil units, heat pumps, boilers, chillers, heat exchangers, pumps, etc., are described in the following sections.

DOI: 10.1201/9781003289326-6

Heating and Cooling of Air Through Coils

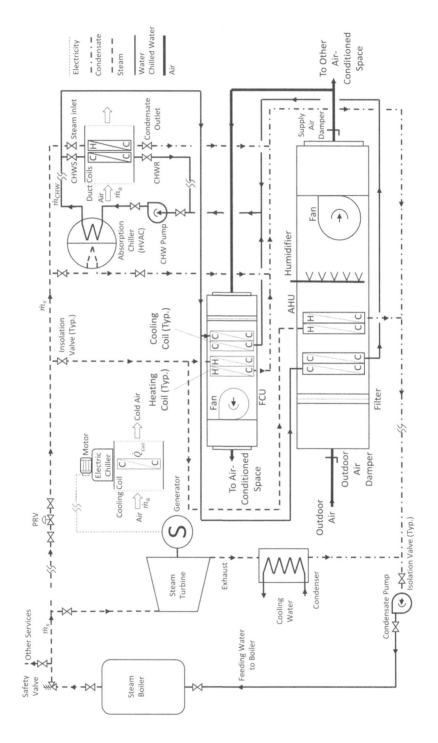

FIGURE 6.1 A coil fluid piping system.

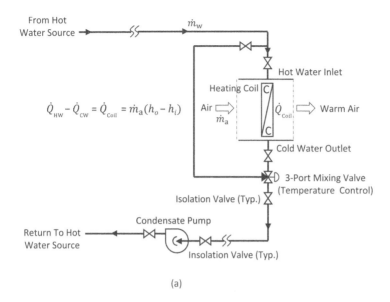

(a)

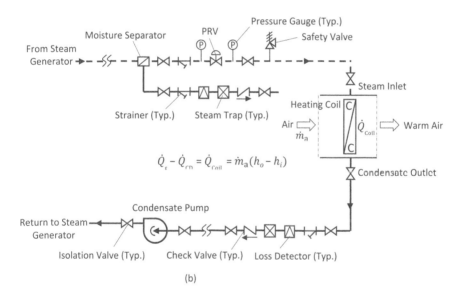

(b)

FIGURE 6.2 Piping systems connecting to the heating coil. (a) The hot water heating system. (b) The steam heating system.

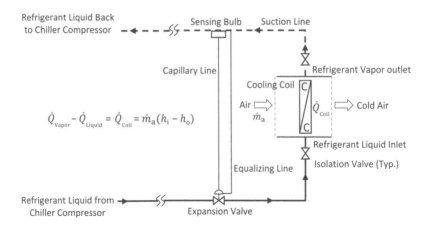

(a)

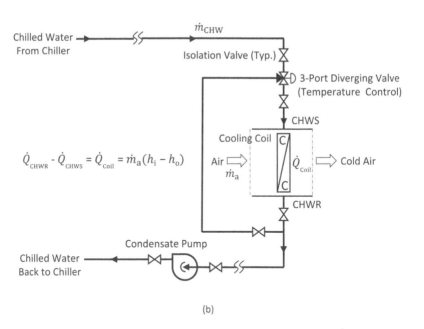

(b)

FIGURE 6.3 Piping systems connected to the cooling coil. (a) The refrigerant system. (b) The chilled water system.

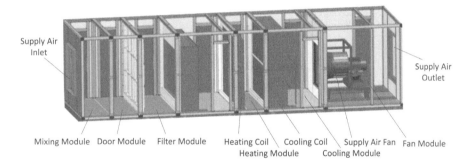

Supply Air
Inlet

Supply Air
Outlet

Mixing Module   Door Module   Filter Module   Heating Coil   Cooling Coil   Supply Air Fan   Fan Module
                                                Heating Module   Cooling Module

FIGURE 6.4   Air handling unit. (Courtesy of Marlo Heat Transfer Solutions, 2022.)

6.4.1 AIR HANDLING UNIT

An air handling unit (AHU), also called an air handler, is a piece of equipment that works as a hub to circulate air and supply conditioned air to the air-conditioned space. Generally, the AHU is a large metal box that houses devices, such as fans, filters, dampers, coils, etc. Figure 6.4 shows a typical AHU.

6.4.2 FAN COIL UNIT

A fan coil unit (FCU), as the name implies, is composed of a fan and coils in a metal cabinet. The coils installed in the cabinet can be a heating coil, a cooling coil, or both. The FCU is a prefabricated device that can operate as a standalone air conditioner or work in a sub-system in a large air conditioning system. Accessories such as a filter and a damper may also be enclosed in the cabinet. Figure 6.5 shows a typical FCU and a schematic of the FCUs in the air conditioning system.

6.4.3 BOILERS

A boiler is a pressure vessel used to transfer heat normally produced by fuel combustion to fluid flowing inside the boiler tubes and generate hot water or steam output. Boilers can be classified into the following categories:

- Purpose of application: There are power boilers and heating boilers. The power boiler is used in power plants to work with a steam turbine unit to generate electricity. The heating boiler, also called an industrial boiler or commercial boiler, works for process heating or home heating.
- Form of fluid output: There are steam boilers and hot water boilers. The steam boiler can be used for power generation and heating. The hot water boiler is commonly used for heating.
- Pressure: Table 6.1 lists the boiler pressure and temperature standards from the ASME Boiler and Pressure Vessel Code (BPVD).

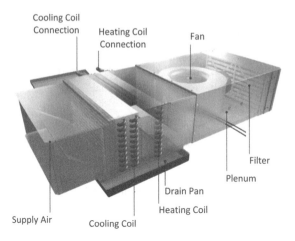

(a)

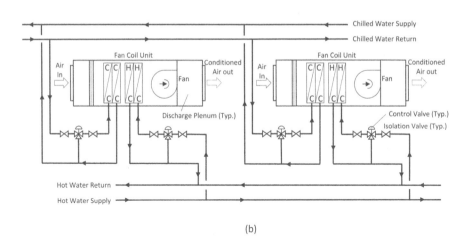

(b)

FIGURE 6.5 Fan coil unit and FCUs in an air conditioning system. (a) Fan coil unit. (b) FCUS in an air conditioning system. (Courtesy of Price Industries, Ltd. 2022.)

The boiler used in a coil fluid system, therefore, can be a steam or hot water boiler. Commonly, it is the hot water boiler.

6.4.3.1 Steam Boilers

The steam boiler generates steam. The steam output can be saturated steam (wet steam) or superheated steam (dry steam). Generally, superheated steam is used for power generation and saturated steam is used for heating. The superheated steam

TABLE 6.1
Boiler Pressure and Temperature Standards

Boiler Function	Steam	Water
Section I, Power Boilers ("S" Stamp)	Vapor Pressure > 103 kPa	Wate Pressure > 1103 kPa and/or Temperature > 121°C
Section IV, Steam Boilers ("H" Stamp)	Vapor Pressure ≤ 103 kpa	Water Pressure ≤ 1103 kPa and/or Temperature ≤ 1103 kPa

produced from a heat exchange (superheater) in the boiler is supplied to a steam turbine to generate electricity. For heating purposes, the saturated steam from the boiler is sent directly to the downstream devices or users, such as heating coils. The return fluid to the steam boiler must be condensated. Figure 6.6(a) and (b) show typical steam boilers for power generation and heating, respectively.

6.4.3.2 Hot Water Boilers
A hot water boiler doesn't boil water. The hot water boiler receives cold water in and turns hot water out. Hot water is circulated in the piping system to release sensible heat in downstream devices, such as heating coils. The return fluid to the boiler is cold water. Figure 6.7 shows typical hot water heating boilers in fire tube and water tube, respectively.

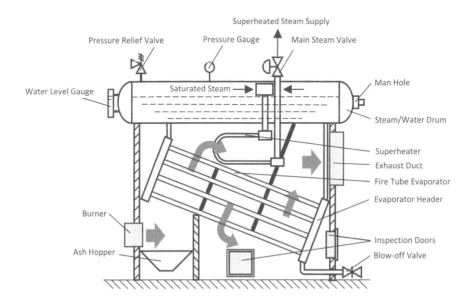

(a)

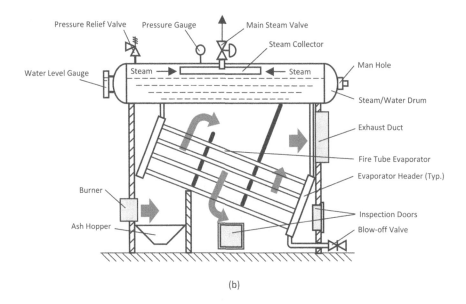

(b)

FIGURE 6.6 Steam boilers. (a) Steam boiler for power generation. (b) Steam boiler for heating.

6.4.4 CHILLERS

A chiller is a machine that generates chilled water flowing inside a cooling coil to cool air passing through the coil. Generally, there are two types of chillers: electric chillers, also called vapor-compression chillers, and absorption chillers, also called heat addition chillers. The electric chiller consumes electricity to drive an electric compressor for circulating refrigerant in the chiller to generate chilled water. The absorption chiller uses heat to generate chilled water. Usually, the absorption chiller has a larger size and heavier weight than an electric chiller at the same chiller capacity. The absorption chiller is preferable because it is driven by a diversity of heating sources and can save a large quantity of electricity.

6.4.4.1 Electric Chiller

An electric chiller is mainly composed of four devices: an electric compressor, a condenser, an expansion valve, and an evaporator. The electric chiller works on a refrigeration cycle. Prefilled refrigerant, such as R134a or R410a, circulates in the chiller to cool return chilled water passing through the evaporator coil. The liquid refrigerant takes heat changes to vapor. The vapor refrigerant is compressed in the electric compressor and becomes vapor with higher pressure and higher temperature. In the condenser coil, the vapor releases heat to the ambient air and changes to liquid with a low temperature. After the expansion valve, the refrigerant

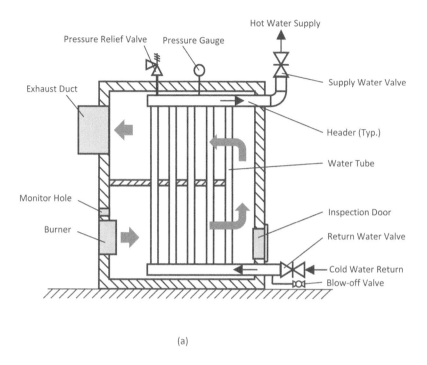

(a)

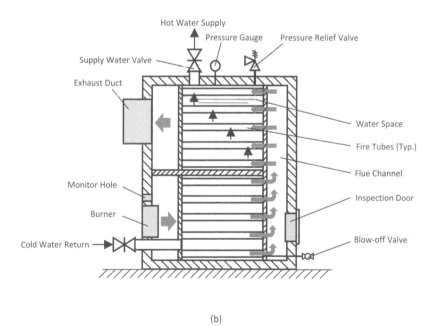

(b)

FIGURE 6.7   Hot water heating boilers. (a) Fire tube boiler. (b) Water tube boiler.

liquid is at the lower temperature and lower pressure condition. In the evaporator, refrigerant is sprayed on the coil surface and vaporized after taking heat from the chilled water flowing inside the coil. Then, the refrigerant vapor flows to the electric compressor and repeats the above processes. Figure 6.8(a) shows a typical electric chiller with a rotary screw compressor, and Figure 6.8(b) shows the working diagram of an electric chiller.

(a)

Water Cooling Condenser — Cooling Water Out
— Cooling Water In

Liquid Refrigerant ↓   Vapor Refrigerant (High Temperature & Pressure)

Expansion Valve ⋈   Electric Compressor

Vapor Refrigerant (Low Temperature & Pressure)

Chilled Water Generator (Evaporator) — Chilled Water Out
— Chilled Water In

Liquid Refrigerant (Low Temperature & Pressure)

(b)

FIGURE 6.8 Electric chiller and its working diagram. (a) Electric chiller with a rotary screw compressor. (b) The electric chiller working diagram. (Courtesy of Broad USA, Inc. 2020.)

6.4.4.2   Absorption Chiller

An absorption chiller is mainly composed of a generator, an absorber, a condenser, an expansion valve, an evaporator, and a few pumps. The absorption chiller works on the refrigeration cycle plus an organic Rankine cycle (ORC) in liquid. In the absorption chiller, the electric compressor in the electric chiller is replaced by the unit of the generator and the absorber. Figure 6.9(a) and (b) shows a system diagram of the absorption chiller and a typical absorption chiller, respectively. In operation, the prefilled liquid mixture of water and lithium bromide (H_2O-LiBr) is heated in the generator. The heat source can be from steam, hot water, wasted heat, solar thermal, ocean thermal, or geothermal, etc. Water in the mixture is vaporized and the lithium bromide stays in a liquid state since LiBr has a higher boiling point than H_2O. The vapor rises and goes to the condenser. In the condenser, vapor is cooled and becomes liquid. After the condenser, water goes to the expansion valve. Through the expansion valve, the water is at a lower pressure and temperature. In the evaporator, water is sprayed on the coil surface, which takes heat from the return chilled water flowing inside the coil. Then, the water changes to vapor. The vapor is absorbed by LiBr in the absorber to form an H_2O-LiBr mixture. The H_2O-LiBr mixture is pumped to the generator to repeat the above processes. In the absorption chiller, water functions as a refrigerant and LiBr is used as a transport medium. Figure 6.9(c) shows a schematic of the working diagram of the absorption chiller.

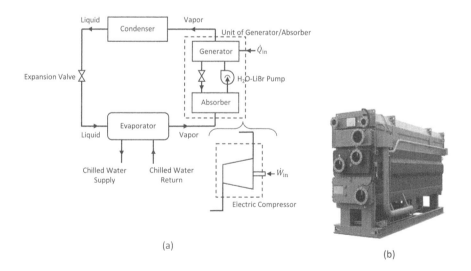

(a)

(b)

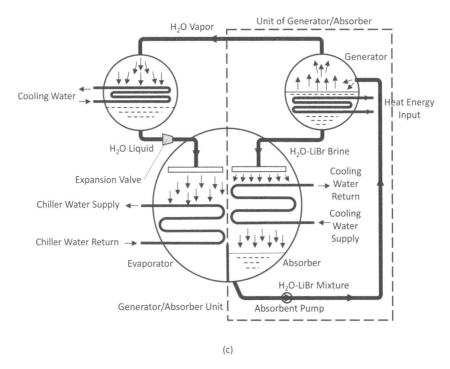

(c)

FIGURE 6.9 Absorption chiller and its working diagram. (a) The system diagram of the absorption chiller. (b) Absorption chiller. (c) The working diagram of the absorption chiller. (Courtesy of Broad USA, Inc. 2020.)

6.4.5 HEAT PUMP

A heat pump typically consists of an evaporator, a condenser, a compressor, a bi-directional expansion valve, and a reverse valve. The heat pump works in the split air conditioning type. The heat pump is a machine that can function as an air conditioner for cooling and as a heater for heating. The function change between cooling and heating is accomplished by the reverse valve. By switching the reverse valve, the flow direction of the refrigerant in the heat pump is reversed. In cooling mode, the indoor coil works as the evaporator to take heat from the room air and the outdoor coil works as the condenser to discharge heat to the ambient air. In heating mode, the outdoor coil works as the evaporator to absorb heat from the ambient air and the indoor coil works as the condenser to release heat to the room. Figure 6.10(a) and (b) shows schematics of the heat pump working diagrams in heating and cooling mode, respectively.

6.4.6 HEAT EXCHANGERS

A heat exchanger is a piece of mechanical equipment that transfers heat between two unmixed fluids flowing in separate loops. They are widely used in engineering practices. In general, heat exchangers have three basic types: pipe, shell and tube, and plate.

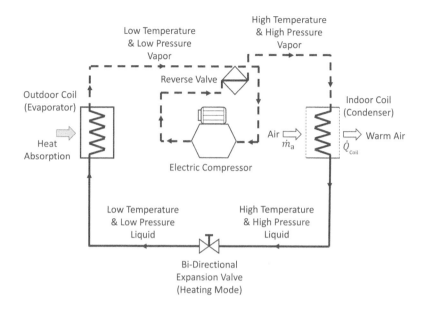

(a)

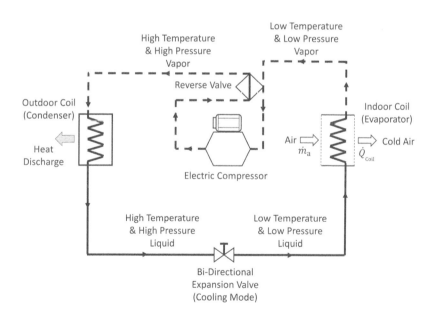

(b)

FIGURE 6.10 Heat pump working diagrams. (a) Heating mode. (b) Cooling mode.

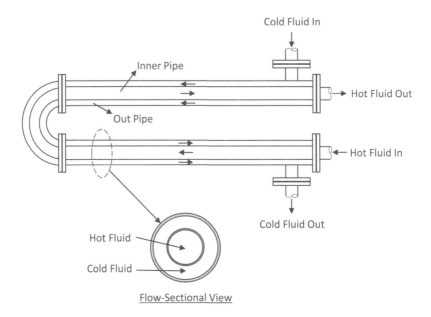

Flow-Sectional View

FIGURE 6.11   A schematic of the pipe heat exchanger.

6.4.6.1   Pipe Type

A pipe heat exchanger, also called double-pipe heat exchanger, is the simplest type of heat exchanger. The exchanger has two pipes. A smaller pipe is held concentrically inside a larger pipe. The surface of the inner pipe acts as a heat transfer boundary where one fluid flows inside the inner pipe and another fluid flows around it in the outer pipe. The heat transfer process occurs between two fluids through the inner pipe wall. The flow sectional area of the pipe heat exchanger is an annulus shape. In engineering applications, the pipe heat exchanger is popular because of its simplicity. Figure 6.11 shows a schematic of the pipe heat exchanger.

6.4.6.2   Shell and Tube Type

A shell and tube heat exchanger consists of four major components: a front header, a rear header, a tube bundle, and a shell. The tube bundle, including tubes, tube sheets, baffles, and tie rods, holds the tubes together. The shell contains the tube bundle in the exchanger. Heat transfer between the two fluids is through the tube surface in the shell. One fluid flowing inside the tubes enters in the front header and moves through the multiple tubes to the rear header. The fluid leaves the exchanger in the rear header or turns back to the front header. Another fluid flows over the tubes in the shell side. Commonly, there are baffles in the shell directing the flow to avoid the fluid taking a shortcut through the shell and have a counter-current flow or cross flow between two fluids. The shell and tube heat exchanger is the most used in engineering practice due to the wide application range of pressure and temperature. Figure 6.12(a) and (b) shows a typical shell and tube heat exchanger and its working diagram of flowing in and out in one header and separate headers, respectively.

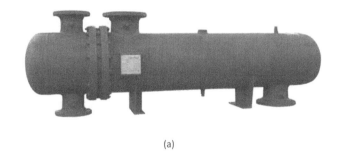

(a)

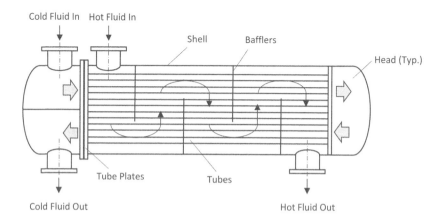

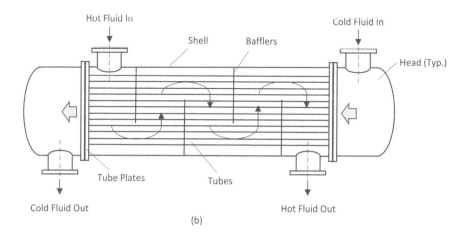

(b)

FIGURE 6.12 Shell and tube heat exchanger and its working diagrams. Shell and tube heat exchanger. (b) The working diagram of the shell and tube heat exchanger. (Courtesy of Marlo Heat Transfer Solutions, 2022.)

6.4.6.3 Plate Type

A plate heat exchanger consists of a series of parallel plates attached side by side to form a series of fluid flow channels. Inlet and outlet holes at the corners of the plates allow hot and cold fluids through alternating plate channels so that one side of a plate is in contact with hot fluid and another side of the plate in contact with cold fluid. The plates are made of steel with good thermal conductivity to transfer heat between two fluids efficiently. The cold and hot fluids flow in a counter flow type. The plate heat exchangers are widely used in engineering practice because of the features of compact and higher heat transfer efficiency. Figure 6.13(a) and (b) shows a typical plate heat exchanger and its working diagram, respectively.

6.4.6.4 Pinch Point Temperature

A pinch point (PP) is a location of the minimum temperature difference between the hot and cold fluids in the heat exchanger. The minimum temperature difference is called the pinch point temperature T_{pp}. The smaller the T_{pp} is, the higher the heat transfer rate of the heat changer will be. However, the more heat transfer surface

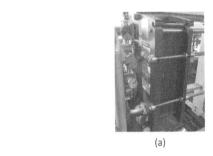

(a)

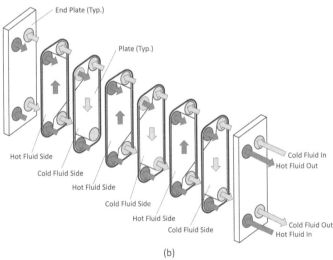

(b)

FIGURE 6.13 Plate heat exchanger and its working diagram. (a) Plate heat exchanger. (b) The working diagram of the plate heat exchanger.

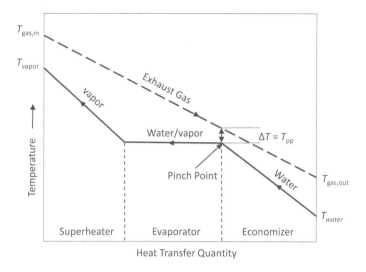

FIGURE 1.3 Pinch point temperature in a heat recovery steam generator.

area of the heat exchanger may need. Figure 1.3 shows a typical diagram of the flow temperatures and heat transfer quantity in a heat recovery steam generator (HRSG). In the diagram, T_{pp} is shown. In the HRSG operation, attention must be paid if the gas temperature at the pinch point is lower than the vapor-saturated temperature of the gas, because vapor in the gas will be condensed to water. The water may play as a corrosive medium to damage the metal surface of the HRSG. The pinch point temperature, therefore, is an important parameter in heat exchanger design and operation that affects the size and thermodynamic performance of the heat exchanger. Commonly the T_{pp} is kept at about 15°C.

6.4.7 Pumps

Pumps are vitally important in forced flow piping systems. The pump provides energy to fluid flowing in the piping system and transports the fluid from one location to the destination. There are various kinds of pumps used in heating and cooling fluid piping systems, such as centrifugal, rotary, reciprocating, and so on. Among them, the centrifugal pump is widely used in engineering practice.

6.4.7.1 Centrifugal Pump

A centrifugal pump is predominantly composed of an impeller, a casing, a shaft, bearings, and a coupling as shown in Figure 6.14(a). The pump converts rotational energy from a motor to the pump shaft and then forces the fluid in motion. The fluid axially enters through the eye of the casing and is discharged through the impeller. The impeller is a rotor used to increase the kinetic energy and pressure of the fluid flowing out the pump. The casing is a shape of volute acting as a pressure containment vessel and directing the fluid flow in and out of the pump. The coupling is an element used to connect the motor shaft for power transfer to the pump shaft. The

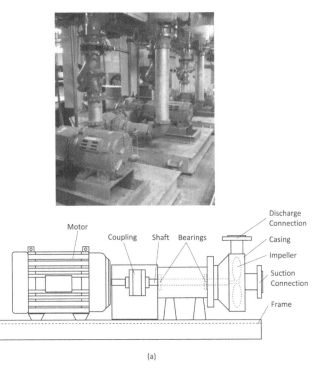

(a)

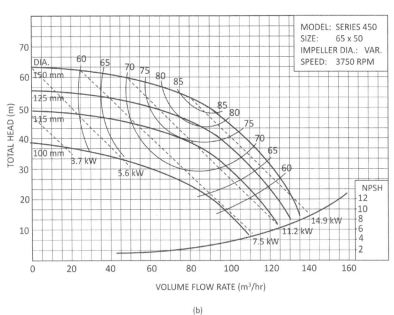

(b)

FIGURE 6.14 Centrifugal pump and the pump performance curve. (a) Centrifugal pump. (b) The performance curve of the centrifugal pump.

shaft is a mechanical component for transmitting torque from the motor to the impeller mounted on it. The bearings constrain the relative motion of the shaft and reduce friction between the rotating shaft and the stator.

The performance of a centrifugal pump is characterized by the volume flow rate (m³/s) and the pump discharge head (m). Figure 6.14(b) shows a typical performance curve of the centrifugal pump. In operation, a pump will perform along its performance curve. The higher the volume flow rate is, the lower the pump discharge head will be and vice versa. The selected pump in application needs to meet the required flow rate and discharge head in the piping system. The designed working condition of the pump should be close to the maximum efficiency point called the best efficiency point (BEP).

When pumping fluid, it is possible to have the local pressure P in the pump falling below the vapor pressure of the fluid P_v, which is the saturation pressure P_{sat} of the fluid corresponding to the fluid temperature. When the fluid pressure in the pump is $P < P_v$, vapor-filled bubbles in the fluid called cavitation bubbles may appear. The cavitation bubbles will cause noise, vibration, lower efficiency, and even damage the impeller blades. In pump operation, cavitation typically occurs on the suction side of the pump where the pressure is usually the lowest. Therefore, $P > P_v$ is required on the suction side of the pump to avoid cavitation. The criterion of measuring if cavitation occurs is called net positive suction head (NPSH). NPSH is expressed as the difference between the pump inlet stagnation pressure head and the vapor pressure head,

$$\text{NPSH} = \left(\frac{P}{\rho g} + \frac{V^2}{2g} \right)_{\text{pump inlet}} - \frac{P_v}{\rho g} \qquad (6.1)$$

To ensure cavitation is not occurring, the available $\text{NPSH}_{\text{available}}$ of the pump in operation must be greater than $\text{NPSH}_{\text{required}}$ provided from the pump manufacturer as illustrated in Figure 6.15. The curve intersect of $\text{NPSH}_{\text{available}}$ and $\text{NPSH}_{\text{required}}$ indicates the maximum volume flow rate delivered by a pump without cavitation in

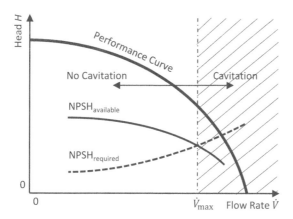

FIGURE 6.15 $\text{NPSH}_{\text{available}}$ and $\text{NPSH}_{\text{required}}$ of a pump.

operation. The NPSH$_{available}$ of a pump varies not only with the flow rate, but also with the liquid temperature, since P_v is the function of temperature.

Example 6.1

A centrifugal pump is used to pump water at a temperature of 25°C from a reservoir whose surface is 4.0 m above the centerline of the pump inlet. The piping system from the reservoir to the pump is new and 20 m long with a few devices as shown. The pipe material is steel and the internal diameter of pipe is 100 mm. If the volume flow rate in the pipe is 1.5 m³/min and the roughness is 0.046 mm, determine *(a)* the NPSH$_{available}$ of the pump and *(b)* if cavitation will occur at the pump inlet based on the pump operation curve shown in Figure 6.14(b).

SOLUTION

Referring to Appendix A.5 Properties of Steam and Compressed Water, the following properties are found at the given water temperature $T = 25°C$,

$$\rho = 997 \frac{kg}{m^3}$$

$$\mu = 0.891 \times 10^{-3} \frac{kg}{m.s}$$

$$P_v = 3.169 \text{ kPa}$$

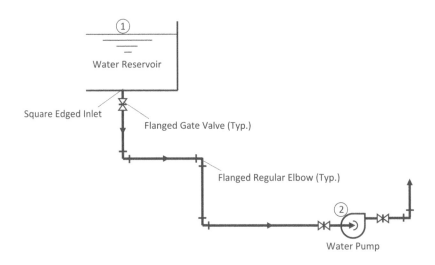

FIGURE Example 6.1

Using Equation (2.47) (see Chapter 2) and flow locations from the reservoir water surface to the pump inlet, the extended Bernoulli equation becomes

$$\frac{P_1}{\rho g} + \frac{1}{2}\frac{V_1^2}{g} + z_1 = \frac{P_2}{\rho g} + \frac{1}{2}\frac{V_2^2}{g} + z_2 + H_{loss}$$

Since the water velocity V_1 is zero and the reservoir water surface is exposed to atmospheric pressure. Then, the pump inlet pressure head is:

$$\frac{P_2}{\rho g} = \frac{P_{atm}}{\rho g} - \frac{1}{2}\frac{v_2^2}{g} + (z_1 - z_2) - H_{loss}$$

or

$$\frac{P_2}{\rho g} + \frac{1}{2}\frac{v_2^2}{g} = \frac{P_{atm}}{\rho g} + (z_1 - z_2) - H_{loss}$$

(a) Substituting above expression into Equation (6.1), the NPSH$_{available}$ at the pump inlet is

$$NPSH_{available} = \frac{P_{atm} - P_v}{\rho g} - \frac{1}{2}\frac{v_2^2}{g} + (z_1 - z_2) - H_{loss}$$

where

$$H_{loss} = \left(f\frac{L}{D} + \sum K_L \right)\frac{V^2}{2g}$$

In the calculation, $V = V_2$, i.e.,

$$V = \frac{\dot{V}}{A} = \frac{4\dot{V}}{\pi D^2} = \frac{4\left(1.5\frac{m^3}{min}\big/60\frac{s}{min}\right)}{\pi(0.1\,m)^2} = 3.1831\,\frac{m}{s}$$

The Reynolds number, therefore, is

$$Re = \frac{\rho V D}{\mu} = \frac{\left(997\frac{kg}{m^3}\right)\left(3.1831\frac{m}{s}\right)(0.1\,m)}{0.891 \times 10^{-3}\frac{kg}{m.s}}$$

$$= 3.5618 \times 10^5 > 4{,}000\;(\text{Turbulent flow})$$

Referring to Appendix A.6 The Moody Chart, friction factor is found

$$f = 0.018$$

based on the relative roughness, $\varepsilon/D = 4.6 \times 10^{-4}$.
From the problem sketch, the sum of the minor loss coefficients is

$$\sum K_L = 0.5 + 3 \times 0.39 + 0.35 = 2.02$$

Substituting the values of f, D, L, and ΣK_L into H_{loss} along with the given properties, the NPSH$_{available}$ with the given volume flow rate is determined to be:

$$\text{NPSH}_{\text{available}} = \frac{(101{,}325 - 3{,}169)\frac{N}{m^2}}{\left(997\ \frac{kg}{m^3}\right)\left(9.81\frac{m}{s^2}\right)}\left(\frac{kg \cdot \frac{m}{s^2}}{N}\right) + 4\ m$$

$$-\left(0.018\frac{20\ m}{0.1\ m} + 2.02 - 1\right)\frac{\left(3.1831\ \frac{m}{s}\right)^2}{2\left(9.81\ \frac{m}{s^2}\right)}$$

$$= 11.65\ m$$

(b) From Figure 6.14(b), the $\text{NPSH}_{\text{required}}$ is read 3.2 m at the volume flow rate 1.5 m³/min = 90 m³/hr, the $\text{NPSH}_{\text{available}}$ at the pump inlet is much bigger than the $\text{NPSH}_{\text{required}}$,

11.65 m $\text{NPSH}_{\text{available}}$ > 4.52 m $\text{NPSH}_{\text{required}}$

Therefore, cavitation will not happen at the pump inlet at the given water temperature and volume flow rate.

6.4.7.2 Pump Arrangement

Pumps can be arranged in serial or parallel to increase the flow head or rate in the piping system. When two or more pumps are connected in series, the overall pump performance curve is the summation of the pump heads at the same pump flow rate as shown in Figure 6.16(a). When two or more pumps are connected in parallel, the overall pump performance curve is the summation of the pump flow rates at the same pump head as shown in Figure 6.16(b). The volume flow rate of each pump in series and the head of each pump in parallel should be the same or close. Otherwise, pumps in serial or parallel with different sizes may cause operation problems. The smaller size pump will restrain the expected overall flow rate or the pump head. An arrangement of one pump much larger than another as shown in Figure 6.16(b) should be avoided.

6.4.8 Valves

In piping systems, valves are important components and widely used to keep or adjust fluid flow at a proper rate. The valves most commonly used in heating and cooling fluid systems are gate, globe, check, ball, etc. Some of them can be used to adjust the flow rate and some of them can only be used fully open or fully closed to isolate the flow. They are called control valves and isolation valves, respectively. Table 6.2 shows a table listing the valves service for control or insolation.

6.4.8.1 Control Valves

A control valve in a piping system is normally used to adjust flow, pressure, or temperature. The control mechanism of the control valve is to vary the flow passage in the valve. To get a lower flow rate, the flow passage of the valve opens smaller.

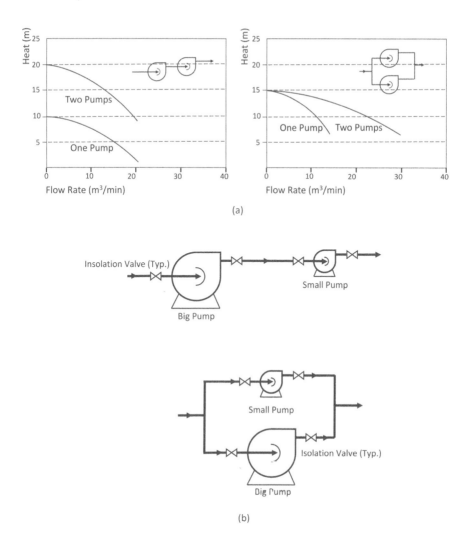

FIGURE 6.16 Pump performance curves and arrangement. (a) Performance curves of pumps in series and parallel. (b) Arrangement of different size pumps (should be avoided).

Otherwise, to get a higher flow rate, the flow passage of the valves opens larger. The control of fluid pressure and temperature is generally accompanied by flow rate adjustment too.

6.4.8.1.1 Pressure Control
A pressure control valve assists in a variety of functions, such as keeping the pressure of a piping system safely below the upper limit or maintaining a constant pressure specified in system operation. The pressure control valves can be categorized in their control mechanisms as direct-operated, pilot-operated, and controller operated. Figure 6.17 shows these types of pressure control valves.

TABLE 6.2
Valve Service

Type	Isolation (Fully Open/Closed)	Throttling (Flow Adjustment)	Pressure Relief (Safety)	Unidirectional (One-Way)
		Service/Function		
Ball	Yes	Note 1	No	No
Butterfly	Yes	Yes	No	No
Plug	Yes	Note 1	No	No
Gate	Yes	No	No	No
Globe	Yes	Yes	No	No
Check	Note 2	No	No	Yes
Pressure Relief	No	No	Yes	No

Notes:

1. Some ball and plug valves designed by some manufacturers are suitable for throttling service. Contact the manufacturers for these kinds of valves.
2. Stop-check valves can be used as stop or isolation valves. Contact the manufacturers for this kind of valve.

- Direct Operated

Both spring-loaded and diaphragm-loaded pressure control valves as shown in Figure 6.17(a) are direct operated pressure-reducing valves. The valves are used to automatically control the downstream pressure at a set pressure. A spring-loaded valve simply uses a spring against a diaphragm to move the valve plug up or down to adjust the flow passage in the valve. As the downstream fluid pressure is higher than the set pressure that corresponds to the set spring position, the outlet fluid pressure

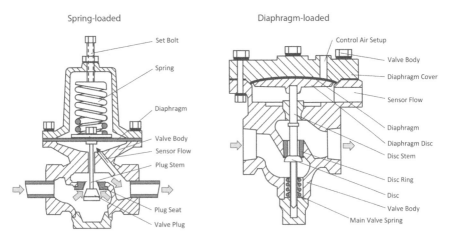

(a)

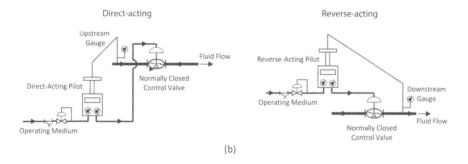

(b)

(c)

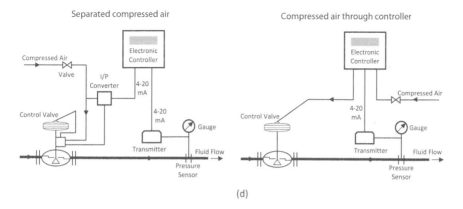

(d)

FIGURE 6.17 Pressure control valves. (a) Directed operated. (b) Pilot operated. (c) Pneumatic diagram pressure valve with the pilot operated reverse-acting control. (d) Controller operated.

under the diaphragm overcomes the spring force and pulls the valve plug upward to the seat. Therefore, the flow passage in the valve becomes narrow, and less flow can go through the valve. As a result, the downstream fluid pressure decreases. Alternatively, as the downstream pressure is lower than the set pressure, the spring pushes the diaphragm down to move the valve plug away from the seat so that the flow passage opens wider. The downstream fluid pressure, therefore, increases. The specified downstream pressure can be manually set up and adjusted by turning the adjusting screw. The diaphragm-loaded valve operates in the same working principle as that of the spring-loaded valve, except that the specified downstream pressure is set up and adjusted by the control air on the top of the diaphragm.

- Pilot Operated

A pilot-operated pressure control valve typically consists of a control valve, a pressure sensor, and a control pilot as shown in Figure 6.17(b). The pressure sensor measures the pressure and transmits it to the control pilot. The control pilot sets the pressure to admit an appropriate operating medium flow toward the control valve. The flow of the operating medium is used to manipulate the valve plug position so that the flow passage can be adjusted. When the upstream fluid pressure is higher than the set pressure, the direct-acting control pilot allows the higher-operating medium flow to the control valve, which enlarges the flow passage in the control valve. As a result, the upstream pressure decreases. When the upstream fluid pressure is lower than the set pressure, the direct-acting pilot acts oppositely to narrow the flow passage in the control valve so that the upstream pressure increases toward the set pressure. When the downstream fluid pressure is higher than the set pressure, the reverse-acting control pilot works to restrict the operating medium flow to the control valve, which will decrease the flow passage in the control valve so that the downstream fluid pressure reduces. When the downstream fluid pressure is lower than the set pressure, the reverse-acting control pilot acts oppositely to enlarge the flow passage in the control valve so that the downstream fluid pressure increases toward the set pressure. Figure 6.17(c) shows a pneumatic diaphragm pressure valve with the pilot-operated reverse-acting control.

- Controller Operated

A controller-operated pressure control valve typically consists of a control valve, a pressure sensor, a transmitter, and an electronic controller as shown in Figure 6.17(d). The pressure sensor measures the fluid pressure. The transmitter converts the pressure to a corresponding signal and sends the signal to the electronic controller. The electronic controller then evaluates the signal with the set pressure to allow a proper compressed airflow to the actuator. The actuator mechanically moves the control valve stem up and down to adjust the flow passage in the control valve. The compressed air used to manipulate the actuator can go through either the electronic controller or a positioner. The compressed airflow is controlled by the instruction issued from the electronic controller. In engineering applications, the electronic controller is commonly connected to a control panel installed in a remote room to accomplish a remote pressure control. The set pressure can be adjusted on the panel screen. Except using compressed air, the actuator can be manipulated by electric voltage, electric current, or hydraulic fluid.

In engineering practice, industrial plants generally produce the highest pressure steam to cover the needs of all processes. The pressure-reducing valve, therefore, is installed on the pipeline to reduce steam pressure to the lower pressure process. Figure 6.18(a) shows the standard installation of a self-contained pilot-operated pressure control valve, which is also called the steam pressure-reducing station (PRS). In the station, two isolation valves are used to close the flow for maintenance or replacement of the control valve. The strainer is used to remove dirt and particles in steam to prevent the control valve from clogging. The gauges are used for operation monitoring. The self-contained pilot valve is used to adjust the steam flow rate through the main control valve as described previously. In operation, the downstream fluid manipulates the pilot valve, which allows the upstream fluid flows in the chamber to move the piston. The piston controls the main valve disc up or down for adjusting the flow passage in the main valve. Major components of the pressure reduction valve include adjusting the spring (in the pilot), valve spring (in the main valve), diaphragm, piston, and valve seats. The set pressure can be adjusted by rotating the pressure-adjusting nut down or up. Figure 6.18(b) shows a typical self-contained pilot-operated pressure control valve.

6.4.8.1.2 Temperature Control

A temperature control valve (TCV) is used to maintain the fluid temperature at a set temperature. There are two types of TCVs: a thermostatic control valve and an actuated control valve. The thermostatic control valve is a self-contained unit without external power involved. The control system is composed of a sensor, capillary tubing, and an actuator as shown in Figure 6.19(a). The valve is manipulated by the expansion of the pre-filled temperature sensible liquid in the control system.

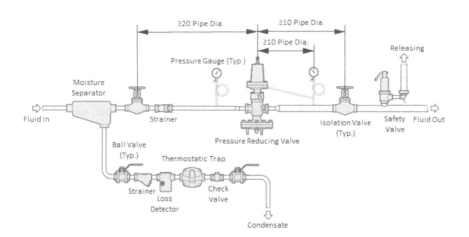

(a)

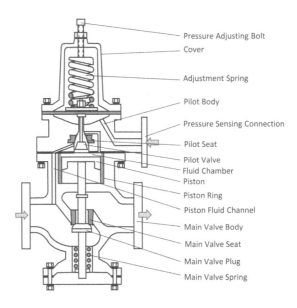

Pressure Adjusting Bolt
Cover

Adjustment Spring

Pilot Body

Pressure Sensing Connection

Pilot Seat

Pilot Valve
Fluid Chamber
Piston

Piston Ring

Piston Fluid Channel

Main Valve Body

Main Valve Seat

Main Valve Plug

Main Valve Spring

Working Mechanism

As downstream pressure exceeds the set pressure, the fluid from the downstream pressure sensor pushes the pilot valve toward the pilot seat, which causes the main valve toward the main valve seat forced by the main valve disc spring. As a results, less flow will go through the main valve to decrease the downstream pressure.

As downstream pressure falls below the set pressure, the adjustment spring pushes the pilot valve away from the pilot seat, which causes more fluid from the front stream to push the piston down and moves the main valve disc away from the main valve seat. As a result, more flow will go through the main valve to increase the downstream pressure.

(b)

FIGURE 6.18 Self-contained pilot-operated pressure control valve. (a) The standard installation of a self-contained pilot-operated pressure control valve. (b) The self-contained pilot. (Courtesy of Spirax Sarco, 2022).

The force created by the liquid expansion in the sensor is transferred via the capillary to the actuator, thereby moving the valve plug toward the seat. The adjustment knob is used to set a temperature by adjusting the liquid space filled in the sensor. For instance, at a higher set temperature, more liquid space is provided and the higher flow fluid temperature can generate a sufficient expansion force to close the valve. The actuated control valve consists of a temperature probe, a transmitter, and a controller. The temperature probe sends a measured temperature to the transmitter. The transmitter sends a signal to the controller. The controller manipulates the flow passage in the TCV through the converter by using an external power source. The actuated control valve is popularly used to control the fluid temperature in heat exchangers as shown in Figure 6.19(b). In application, the TCV controls the temperature of cold fluid exiting the heat exchanger. When the temperature of the exiting

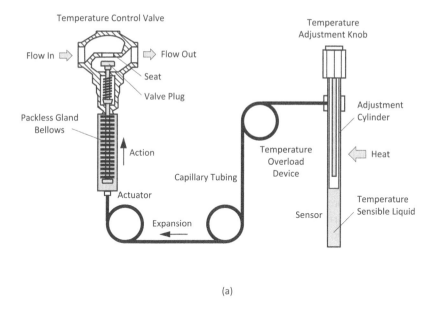

(a)

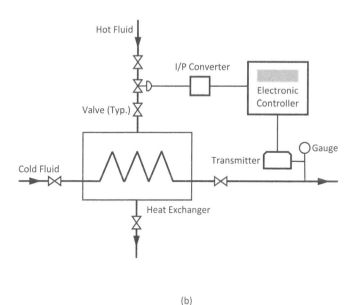

(b)

FIGURE 6.19 Typical temperature control valves. (a) Self-acting. (b) Controller.

fluid is probed higher than the set temperature, the controller issues an instruction actuating the control valve to narrow the flow passage in the valve so that the hot fluid flow is reduced. As a result, the temperature of the cold fluid exiting the heat exchanger decreases back to the set temperature.

Pressure and TCVs may fail in operation due to various factors, for instance, erosion of the seat seals over time, temperatures or pressures out of the designed range, abrasive debris clogging the valve seats, loss of signals, and/or unexpected power loss. There are three valve positions as valve failure occurs in operation to protect the downstream equipment and components. The positions are fail open (FO), fail closed (FC), and fail in the last position (FL).

- FO means the valve will go to a full open position regardless of the valve position before the failure. FO is the option to be used for downstream devices requiring a continuing flow.
- FC means the valve will go to a closed position regardless of the valve position before the failure. FC is the option to be used for preventing downstream devices from a continuing flow.
- FL means the valve will stay in the last or current position when failure occurs.

6.4.8.2   Insolation Valves

An isolation valve, also called an on-off valve, is used either in a full open or full closed position. Isolation valves cannot be used for flow or flow parameter adjustment because their seat and disc configurations will create strong turbulence in the flow and cause a significant head loss, which results in inaccurate adjustments. In normal operations, the isolation valve should remain in the full open position. Under certain circumstances, the valve will be in the full closed position, such as piping system maintenance, valve repair and replacement, or other safety reasons.

6.4.8.2.1   Ball Valve

A ball valve consists of a ball with a flow channel. Turning the valve to align the channel with the pipe, the valve is full open. Turning the valve to block the channel with the pipe wall, the valve is full closed. Figure 6.20(a) shows a typical ball valve.

6.4.8.2.2   Butterfly Valve

A butterfly valve consists of a flat disc that connects with the valve stem. The butterfly valve works like a door to rotate the flat disc for full open or closed. Figure 6.20(b) shows a typical butterfly valve.

6.4.8.2.3   Plug Valve

A plug valve is a quarter-turn valve using a cylindrical or tapered plug with a passage to allow the fluid through the valve. By rotating the plug passage to align the passage with the pipe, the valve is fully open. Turning the valve to block the passage with the pipe wall, the valve is full closed. In many respects, the plug valve is similar in operation to ball valves. However, the plug valve can be used for flow control. Figure 6.20(c) shows a typical plug valve.

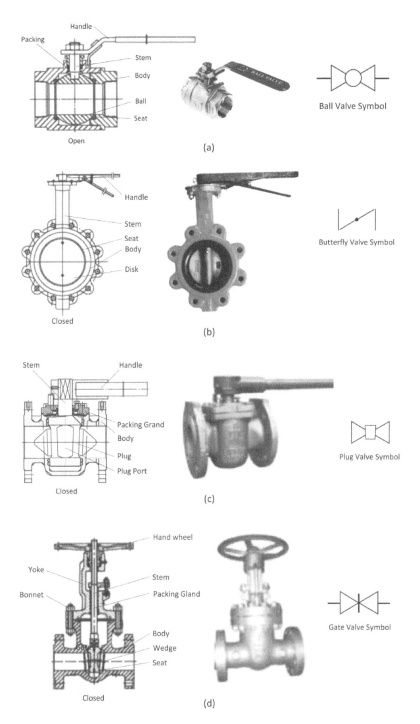

FIGURE 6.20 Typical isolation valves. (a) Ball valve. (b) Butterfly valve. (c) Plug valve. (d) Gate valve. (Courtesy of CWT Valves Industries Inc., 2022.)

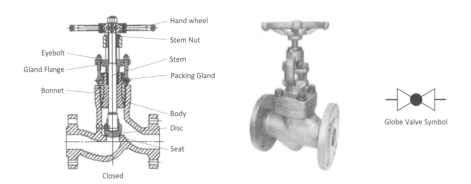

FIGURE 6.21   Globe valve. (Courtesy of CWT Valves Industries Inc., 2022.)

6.4.8.2.4   *Gate Valve*

A gate valve, also known as a sluice valve, has a gate (barrier) in the valve. The valve provides a full flow by lifting the gate out of the fluid path and shuts off the flow by sliding the gate down completely. Figure 6.20(d) shows a typical gate valve.

6.4.8.3   Other Valves

6.4.8.3.1   *Globe Valve*

A globe valve is named for its spherical body shape and can be used for regulating flow in a pipeline. Major components in a globe valve are a movable plug or disc element and a stationary ring seat to control the flow. In the globe valve, the plug is connected to a stem. The stem moves the plug away or to the seat for flow control. Figure 6.21 shows a typical globe valve.

6.4.8.3.2   *Check Valves*

A check valve is a device that allows fluid to flow in only one direction to prevent a backflow in a pipeline. The check has two ports, one inlet and one output. There are two basic types of check valves: swing and lift. A swing check valve consists of a disc that can rotate around a hinge pinned on the valve top to open and close. A lift check valve consists of a disc that can be lifted. Both check valves work on the same mechanism. In normal operations, the fluid flow pushes the disc off the valve seat to open the flow. A reverse flow forces the disc down toward the valve seat to shut off the flow. On the valve, a sign of the flow direction is usually shown to avoid improper installation, causing the check valve to work incorrectly. Figure 6.22 shows a typical swing valve and lift check valve, respectively.

6.4.8.3.3   *Pressure Relief Valve*

A pressure relief valve is installed to protect downstream equipment from overpressure by quickly releasing fluid in the pipeline to reduce the downstream fluid flow pressure. The pressure relief valve mainly consists of a spring, a spring adjuster, and a disc, as shown in Figure 6.23. A spring adjuster is used to set the spring force corresponding to the specified working pressure in the pipeline. As the fluid pressure

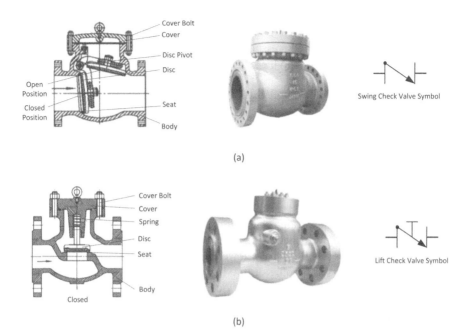

FIGURE 6.22 Typical check valves. (a) Swing check valve. (b) Lift check valve. (Courtesy of CWT Valves Industries Inc., 2022.)

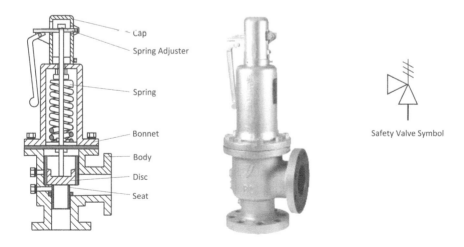

FIGURE 6.23 A typical pressure relief valve. (Courtesy of CWT Valves Industries Inc., 2022.)

in the pipeline is higher than the set pressure, the fluid pushes the disc off the seat. The pressure relief valve opens to allow the fluid out so that the fluid pressure in the pipeline decreases. As the fluid pressure in the pipeline goes back to the set pressure, the spring forces the disc back to the seat and the pressure relief valve is closed.

6.4.8.3.4 Strainer

A strainer is not a valve. The strainer is a pipeline fitting through which debris in the fluid is filtered out and kept in the strainer. The strainer is a necessary component to protect expensive and critical downstream equipment and components, for example, control valves, pumps, flow meters, steam traps, etc., from harmful effects. The strainer can be classified into two main types according to their body configuration: Y-type and basket type.

- Y-Type: The Y-type strainer has a filter leg that connects to the pipeline at a diagonal angle, which gives the strainer a "y" shape and its name. Y-type strainers usually have a lower debris-holding capacity than basket-type strainers. The plug on the bonnet at the end of the leg can be removed to blow debris out. To service the strainer, the bonnet needs to be removed. In application, when a large amount of debris is expected, a blowdown valve can be fitted on the bonnet that enables the strainer to be cleaned conveniently. Y-type strainers can be installed in either horizontal or vertical pipelines. Figure 6.24(a) shows a typical Y-type strainer.
- Basket Type: The basket-type strainer is characterized by a vertically oriented basket. Typically, the basket-type strainer is larger than the Y-type strainer. The debris-holding capacity is also greater than that of the Y-type strainer. The plug on the bottom of the basket can be removed to blow out debris. To service the strainer, the bonnet on the top of the strainer needs to be removed to access the filtering element. Basket-type strainers can only be installed in horizontal pipelines. Figure 6.24(b) shows a typical basket-type strainer.

6.4.9 FANS

A fan is a machine used to create the air flow. The working principle of the fan is similar to that of the pump. The pump is used to move liquid and the fan is used to move air or gas. The major component of the fan is the rotating assembly, also called an impeller, which consists of blades and a hub. In general, fans are driven by electric motors. There are three main types of fans: axial, centrifugal, also called radial, and cross flow, also called tangential. In engineering practice, centrifugal fans are widely used.

6.4.9.1 Centrifugal Fans

A centrifugal fan uses the kinetic energy of the impeller to move the air stream. Air enters the impeller axially through the inlet. The impeller blades swing the air from a smaller to a larger radius, which is collected by a spiral-shaped casing known as volute casing. The gas is discharged out from the casing exit at a higher pressure and velocity.

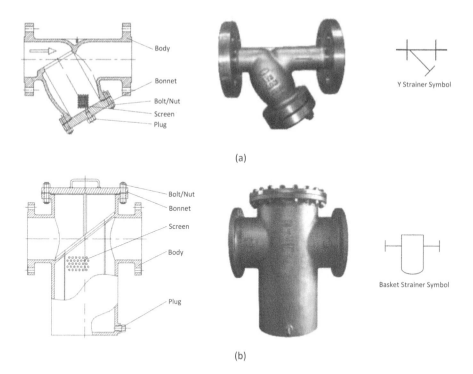

(a)

(b)

FIGURE 6.24 Typical strainers. (a) Y type. (b) Basket type. (Courtesy of CWT Valves Industries Inc., 2022.)

6.4.9.1.1 Fan Blade Types

Fan blades on the hub have four basic shapes: backward curved, straight radial, forward curved, and airfoil as shown in Figure 6.25.

- Backward curve: The blade curves against the direction of the fan wheel rotation. The backward curve blade provides better energy efficiency than others. The fan of backward curved blades can handle gas streams from low

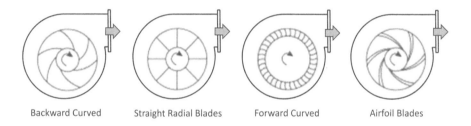

Backward Curved   Straight Radial Blades   Forward Curved   Airfoil Blades

FIGURE 6.25 Fan blade types.

to moderate particulate loading. The fan may have a high range of specific speeds in operation.

- Straight Radial: The blade extends straight out from the center of the hub. The fan of a straight radial blade is typically used for particulate-laden gas streams because the fan is the least sensitive to solid build-up on blades. The fan often generates greater noise on the output side. High speeds, low volumes, and high pressures are common characteristics of the straight radial blade fan.
- Forward curved: The blade curves in the same direction as the fan wheel rotation. The shape of the forward curve makes the blade sensitive to particulates. The fan of the forward curved blade commonly is used in the application of clean air. The forward curved fan can provide lower air flow with a higher static pressure. Forward-curved fans are less efficient than backward-curved fans.
- Airfoil: The blade is shaped like that of an airplane wing. The fan offers the highest efficiency and lower noise. But the fan of airfoil blade is expensive to build. The airfoil fans are typically applied in the application of high static pressure.

6.4.9.1.2   Drive Mechanisms

Fan drives can be classified into two basic types: direct-drive and indirect-drive. Figure 6.26 shows direct-drive and indirect-drive arrangements, respectively.

- Direct-Drive: In direct-drive, the fan motor shaft is directly connected to the fan wheel shaft. The connection is accomplished by a common shaft or a mechanical coupling. Direct-drive eliminates the power transmission loss. The fan wheel rotates at the same speed as that of the motor. In engineering applications, a variable frequency drive (VFD) is commonly applied to alter the motor speed matching the fan speed or adjust the fan speed. The direct-drive fan offers the advantages of lower cost, fewer components involved in assembly, less maintenance, and higher efficiency compared to the belt-drive fan.
- Indirect-Drive: The indirect-drive, also called belt-drive, means that a belt is applied as an intermediary link between a driving motor and a driven fan. More components are involved in indirect-drive. Pulleys are key components mounted on the motor and fan shafts. The belt transmits the rotation energy from the motor to the fan. The fan wheel speed depends upon the diameter ratio of the motor pulley to the fan wheel pulley. Change of the fan speed can be achieved by adjusting the pulley location on the motor shaft or replacing the pulley in different diameters. Indirect-drive offers the feature of flexibility in connection orientation between the fan and the motor.

Table 6.3 shows some features of fan in the direct-drive and the indirect-drive. When equipped with a VFD, direct-drive can offer great flexibility of the speed alternation.

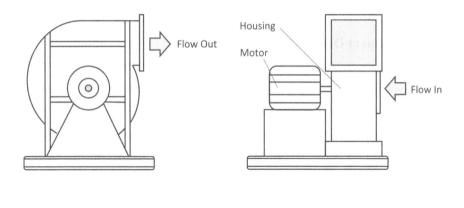

(a)

(b)

FIGURE 6.26 Fan drive types. (a) Direct-drive. (b) Indirect-drive.

6.4.9.2 Fan Laws

Fan laws, also known as affinity laws, are used to express the relationship among performance variables involved in fan operation, such as head, volumetric flow rate, shaft speed, and power consumption. The affinity laws are a set of equations that are useful for determining the impact of extrapolating from a known fan performance to a desired performance, which is valid for both centrifugal fans and pumps:

$$\frac{\dot{V}_2}{\dot{V}_1} = \frac{\omega_2}{\omega_1}\left(\frac{D_2}{D_1}\right)^3 \tag{6.2a}$$

$$\frac{H_2}{H_1} = \left(\frac{\omega_2}{\omega_1}\right)^2\left(\frac{D_2}{D_1}\right)^2 \tag{6.2b}$$

TABLE 6.3

Features of Direct-Drive and Indirect-Drive

Factors in Consideration	Direct-Drive (Shaft Connection)	Indirect-Drive (Belt)
First cost	Generally lower than indirect-drive.	Generally higher than direct-drive.
Range of operating speed	Fan cannot operate below recommended motor speed.	Changing pulleys or the location on the motor shaft can allow the fan to operate in low-speed applications.
Efficiency	Higher efficiency. Less energy loss because of less friction with a shaft connection.	Less efficiency. More energy loss because of more friction from the belt.
Downstream contaminant concerns	None	Some due to belt residue.
Maintenance complexity	Field motor or fan change involves disassembly of integral shaft connection.	Field motor change involves motor, belts, and sheaves.
Vibration	Less after correct alignment in connection of fan and motor.	Additional spinning elements that could be required for balance.
Noise	Louder	More silent
Arrangement with motor	Less flexibility	Greater flexibility

$$\frac{kW_2}{kW_1} = \frac{\rho_2}{\rho_1}\left(\frac{\omega_2}{\omega_1}\right)^3\left(\frac{D_2}{D_1}\right)^5 \tag{6.2c}$$

where ω is shaft rotational speed, D is propeller diameter, ρ is fluid density, H is discharge head, and kW is power consumption. For a known or an existing fan or pump, Equation (6.2) reduces as

$$\frac{\dot{V}_2}{\dot{V}_1} = \frac{\omega_2}{\omega_1} = \frac{\dot{n}_2}{\dot{n}_1} \tag{6.3a}$$

$$\frac{H_2}{H_1} = \left(\frac{\omega_2}{\omega_1}\right)^2 = \left(\frac{\dot{n}_2}{\dot{n}_1}\right)^2 \tag{6.3b}$$

$$\frac{kW_2}{kW_1} = \left(\frac{\omega_2}{\omega_1}\right)^3 = \left(\frac{\dot{n}_2}{\dot{n}_1}\right)^3 \tag{6.3c}$$

where $\dot{n}$ is rpm.

Example 6.2

A fan circulates air in a room. The power consumption is 5 kW to circulate the air at a rate of 60 m³/min. If the air circulation is doubled, determine the power consumption by using the same fan.

SOLUTION

Using Equation (6.3a), the way that the technician can double the volume flow rate by using the same fan is to double the fan speed, i.e. $\dot{n}_2 = 2\dot{n}_1$, i.e.,

$$\frac{\dot{V}_2}{\dot{V}_1} = \frac{2\dot{V}_1}{\dot{V}_1} = \frac{2\dot{n}_1}{\dot{n}_1}$$

Using Equation (6.3c), the required power consumption is

$$kW_2 = \left(\frac{2\dot{n}_1}{\dot{n}_1}\right)^3 kW_1 = \left(\frac{2\dot{V}_1}{\dot{V}_1}\right)^3 kW_1 = 8(5 \text{ kW}) = \mathbf{40\ kW}$$

The power consumption is 8 times bigger to double the air flow rate by using the same fan.

6.4.10 DAMPERS

Dampers are the devices used to vary air flow rate in ductwork by adjusting the flow passage of the dampers. The adjustment of the damper can be manual or mechanical. The manual damper is adjusted by using a handle on the side of the damper. The mechanical damper is adjusted by an actuator powered electrically or pneumatically. There are two basic styles of dampers: parallel blades and opposed blades as shown in Figure 6.27(a). The opposed blades tend to give a more uniform air flow around the centerline of the damper. Figure 6.27(b) shows their typical applications in the air conditioning system.

6.4.10.1 Outdoor Air Damper

An outdoor air damper controls the amount of outside air induced into the air conditioning system. Outdoor air dampers usually are sized to match the dimensions of the outdoor stationary louvers. Outdoor air dampers can be manipulated to open and close, either manually or via motorized means.

6.4.10.2 Return Air Damper

A return air damper, also called a mixing air damper, is used to maintain a constant pressure at the inlet side of the fan, which has the same pressure drop as the combined pressure drop across the louver and damper at the intake air side. Return air dampers usually are the opposed blade dampers and can be interlocked with the outside air dampers.

6.4.10.3 Volume Air Damper

A volume air damper is used to control airflow to individual spaces or zones. Volume air dampers usually are the opposed blade dampers.

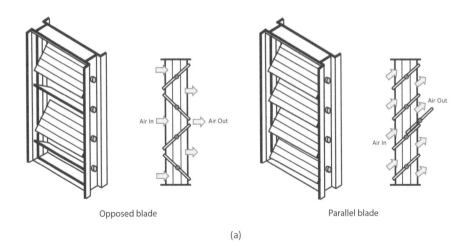

(a)

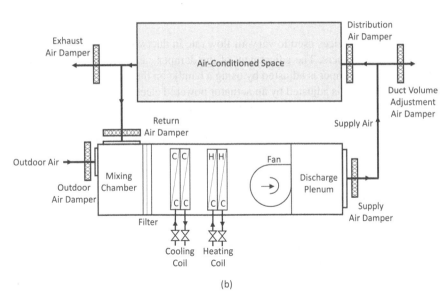

(b)

FIGURE 6.27 Damper types and damper application. (a) Two basic types of dampers. (b) Damper application in an air conditioning system.

7 Energy Sources of Heating and Cooling

In the heating process of air passing through a coil, air obtains heat from hot fluid flowing inside the coil. In the cooling process of air passing through a coil, the air releases heat to cold fluid flowing inside the coil. Heat transfer between the air and the fluid should be balanced with the energy provided from energy sources. There are a variety of energy sources available. Conventional energy sources, such as fuel combustion and electricity, are common. However, gases of nitrogen dioxide (NO_2) and carbon dioxide (CO_2) from fuel combustion discharged with the exhaust bring concerns of air and environment pollution. Alternatively, renewable energy sources, for instance, solar energy, geothermal energy, ocean thermal energy, and wind energy, are becoming attractive. Renewable energies can be used for either heating or electricity generation—except wind energy, which can be used only for electricity generation.

7.1 WORKING FLUIDS IN COILS

There are four types of working fluids in coils used in heating and cooling. They are steam, hot water, chilled water, and refrigerant. The first two fluids are used for heating and last two fluids are used for cooling.

7.1.1 STEAM

Steam can be generated from a steam generator (e.g., either a boiler or a heat exchanger) by a conventional energy source or a renewable energy source. In a coil, steam transfers heat to air passing through the coil. Leaving the coil, steam becomes condensate, which returns to the steam generator. Alternatively, steam can be used as heating energy input in an absorption chiller to generate chilled water. In a steam turbine power plant, steam is used to generate electricity. Electricity can be used to produce heating fluid flowing inside the coil or drive an electric chiller to generate chilled water.

7.1.2 HOT WATER

Similar to steam, hot water can be generated in either a boiler or a heat exchanger by conventional or renewable energy sources. In a coil, hot water transfers heat to the air passing through the coil. Leaving the coil, hot water becomes cold water, which returns to the hot water generator. Alternatively, hot water can be used as heating energy input in an absorption chiller to generate chilled water.

7.1.3 REFRIGERANT

A refrigerant is used as a cooling fluid flowing in a coil to take heat directly from air passing through the coil in a standalone air conditioner, a heat pump, or a DX

DOI: 10.1201/9781003289326-7

cooling system. Alternatively, refrigerant can be used to work in an electric chiller to generate chilled water.

7.1.4 CHILLED WATER

Chilled water used as a cooling fluid flowing inside a coil takes heat from air passing through the coil. Chilled water can be generated from an electric chiller or an absorption chiller. Chilled water circulates in a chilled water pipeline connecting to the chiller and the cooling coil.

7.2 CONVENTIONAL ENERGY SOURCES

7.2.1 FUEL COMBUSTION

When a fuel is in combustion, water in the fuel becomes vapor. If the vapor is discharged out with the combustion exhaust, the heat released from fuel combustion is called the low heating value (LHV) of the fuel. If the vapor is condensed to liquid, the heat released from the fuel combustion is called the high heating value (HHV) of the fuel. The difference between the HHV and the LHV is called latent heat Q_{latent} of water vaporization in fuel combustion, i.e.,

$$Q_{latent} = HHV - LHV$$

To recover the latent heat from the exhaust, additional equipment may be required to convert the vapor to liquid. Normally, latent heat is not recovered in the heat transfer process. Therefore, the heat released from the fuel is LHV. The mass flow rate of steam or hot water generated from a boiler is determined to be

$$\dot{m} = \eta_{boiler} \frac{\dot{G}LHV}{(h_l - h_e)} \tag{7.1}$$

where η_{boiler} is the boiler efficiency, $\dot{G}$ is the fuel consumption rate, and h_l and h_e are the enthalpies of water entering and leaving the boiler, respectively. For a steam boiler, h_l is the enthalpy of the steam leaving the boiler. The required boiler heat rate $\dot{Q}_{boiler}$, therefore, becomes

$$\dot{Q}_{boiler} = \dot{m}(h_l - h_e)$$

If any heat losses in the pipeline and the coil are neglected, the $\dot{Q}_{boiler}$ should be equal to the heat transfer rate of the coil $\dot{Q}_{coil}$, which is the same as the heat rate transferred to air passing through the coil $\dot{Q}_a$,

$$\dot{Q}_{boiler} = \dot{Q}_{coil} = \dot{Q}_a = \dot{m}_a(h_o - h_i)$$

where $\dot{m}_a$ is the mass rate of air passing through the coil, h_o and h_i are the enthalpies of air entering and leaving the coil. Figure 7.1 shows schematics of the hot water and steam pipeline connecting to the heating coil, respectively.

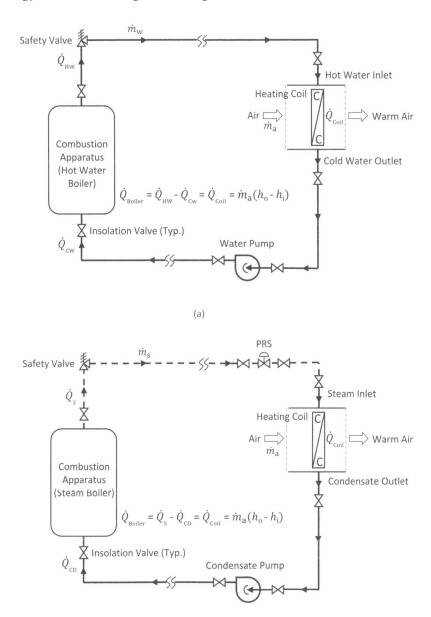

(a)

(b)

FIGURE 7.1 Hot water and steam pipeline connecting to the heating coil. (a) The hot water system. (b) The steam system.

Example 7.1

An airflow at a rate of 73 m³/min in an HVAC duct system enters a heating coil at 15°C T_{db} and 14°C T_{wb}. The air leaves the coil at 22°C T_{wb}. A steady hot water flow from a boiler enters the coil at 82°C and leaves at 76°C, respectively. The waterspecific heat c_p is 4.2 kJ/kg.K, the boiler efficiency is 80%, and the fuel LHV is 42,000 kJ/kg. Neglecting any heat losses in the water piping system and the coil, determine (a) the hot water flow rate (kg/min) from the boiler and (b) the fuel consumption rate (kJ/hr) in the boiler to meet the water and air conditions.

SOLUTION

The states of the air entering and leaving the coil are specified by the given temperatures. Referring to Appendix A.4 Psychrometric Chart at aa Pressure of 1 atm (101.325 kPa), the following properties of the air entering and leaving the coil are found,

$$h_1 = 39.31 \frac{kJ}{kg \ dry \ air}$$

$$v_1 = 0.8263 \frac{m^3}{kg \ dry \ air}$$

and

$$h_2 = 46.61 \frac{kJ}{kg \ dry \ air}$$

(a) The heat transfer rate to the air should be equal to the heat rate supplied from the hot water $\dot{Q}_w$, i.e.,

$$\dot{Q}_a = \dot{Q}_w$$

$$\dot{Q}_a = \dot{m}_a(h_2 - h_1)$$

$$\dot{Q}_w = \dot{m}_w c_p \left(T_i - T_o\right)$$

The air mass flow rate is determined to be

$$\dot{m}_a = \frac{\dot{V}_1}{v_1} = \frac{73 \ \frac{m^3}{min}}{0.8263 \frac{m^3}{kg \ dry \ air}} = 88.35 \ \frac{kg}{min}$$

The heat transfer rate of the air is

$$\dot{Q}_a = \left(88.35 \ \frac{kg}{min}\right)\left(46.61 - 39.31 \ \frac{kJ}{kg}\right) = 644.96 \ \frac{kJ}{min}$$

Then the hot water flowrate from the boiler is

$$\dot{m}_w = \frac{\dot{Q}_w}{c_p \left(T_i - T_o\right)} = \frac{644.96 \ \frac{kJ}{min}}{\left(4.2 \ \frac{kJ}{kg.K}\right)(82 - 76)°C}$$

$$= 25.59 \ \frac{kg}{min}$$

(b) Using Equation (7.1), the fuel consumption rate in the hot water boiler is

$$\dot{G} = \frac{\dot{m}_w (h_o - h_i)_{Boiler}}{\eta_{boiler} LHV}$$

$$= \frac{644.96 \ \dfrac{kJ}{min}}{80\% \left(42,000 \ \dfrac{kJ}{kg}\right)} = 0.0192 \ \frac{kg \ fuel}{min} = 1.1517 \ \frac{kg \ fuel}{hr}$$

7.2.2 Electricity

Electricity is considered one of the most clean and environmentally friendly energy sources. An electric apparatus uses electricity, such as electric boiler or electric heater, to generate steam or hot water for heating, or an electric chiller using refrigerant to generate chilled water for cooling. Figure 7.2 shows schematics of using electricity in heating and cooling. In the figures, the heat engine is an internal combustion engine, such as a diesel engine or a gas turbine.

When using electricity for heating, overall efficiency may be compromised. The overall efficiency of the heating system using hot water or steam in is

$$\eta_{overall.water\_steam} = \eta_{combustion} \eta_{thermal}$$

And the overall efficiency of the heating system using electricity is

$$\eta_{overall.electricity} = \eta_{combuson} \eta_{thermal} \eta_{mechanical} \eta_{generator}$$

The overall efficiency of the heating system using electricity is smaller than that of using hot water or steam. Therefore, using electricity as a heating source may not be economically justified from the view of energy utilization.

Example 7.2

An air heating system uses an electric heater to heat air. Alternatively, a steam heating coil can be used in the system to heat the air. If the fuel combustion efficiency is 90%, the thermal efficiency is 65%, the mechanical efficiency is 70%, and the generator efficiency is 98%, determine the overall efficiencies of the energy utilization in the heating system by using the electric heater and the steam heating coil.

SOLUTION

Steam: $\eta_{overall.steam} = \eta_{combustion} \eta_{thermal}$
$$= 90\% \times 65\% = \mathbf{58.5\%}$$

Electricity: $\eta_{overall.electricity} = \eta_{combustion} \eta_{thermal} \eta_{mechanical} \eta_{generator}$
$$= 90\% \times 65\% \times 70\% \times 98\% = \mathbf{40.1\%}$$

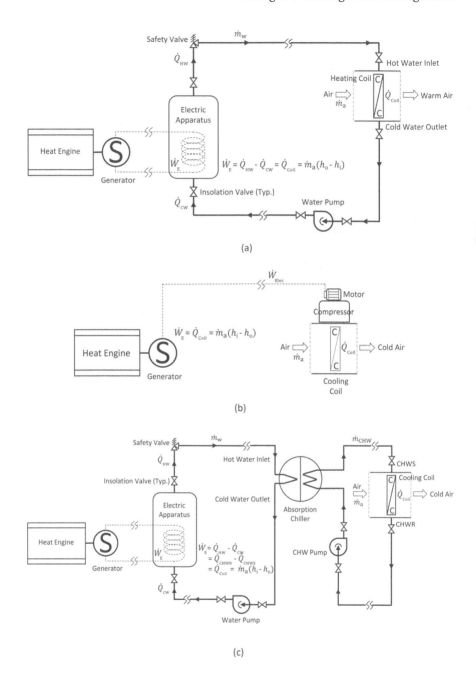

FIGURE 7.2 Schematics of electricity application in heating and cooling. (a) Hot water heating. (b) Refrigerant cooling. (c) Absorption chiller power by hot water generated from electricity.

7.3   WASTE HEAT RECOVERY

Waste heat occurs in almost all thermal machines and processes. Waste heat recovery aims to reuse the energy of the waste heat that otherwise would be disposed of or simply discharged to the environment. The energy recovered from waste heat can be used for the heating and cooling of air passing through coils. Waste heat recovery can not only improve the rate of energy utilization, but also protect the environment.

7.3.1   DIESEL ENGINE JACKET WATER

The diesel engine uses the cooling water through the engine jacket to prevent the engine from overheating in operation. The cooling water is commonly called jacket water which has discharged temperature up to 82°C. The jacket water is usually discharged to the surrounding environment. To recover the jacket water heat for heating and cooling of air passing through coils, jacket water can be pumped directly to the heating coil or the absorption chiller as heating energy to generate chilled water as shown schematically in Figure 7.3(a). Alternatively, jacket water can be directed to a

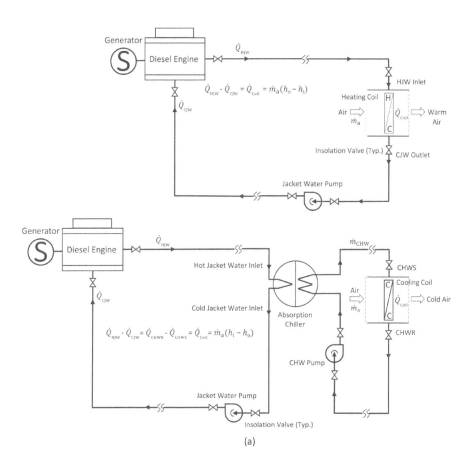

(a)

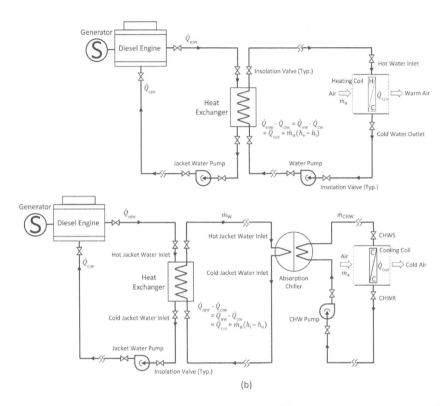

FIGURE 7.3 Heat recovery from engine jacket water. (a) Directly pumped to the heating coil or absorption chiller. (b) An independent hot water loop with a heat exchanger.

heat exchanger to transfer heat to an independent hot water loop as shown schematically in Figure 7.3(b). The application of the independent hot water loop can prevent sediments from the heating coil clogging the jacket water passage in the diesel engine.

7.3.2 Gas Turbine Exhaust

The gas turbine has a lower thermal efficiency. One of the reasons for this is a high exhaust temperature ranging from 315°C to 500°C. To recover exhaust heat and improve the overall thermal efficiency of the gas turbine, the exhaust can be directed to the heating coil or the absorption chiller as heating energy input to generate chilled water as shown schematically in Figure 7.4(a). Alternatively, the exhaust can be directed to a heat exchanger to transfer heat to an independent hot water loop as shown schematically in Figure 7.4(b). The application of the independent hot water loop can prevent sediments from the exhaust clogging the heating coil. In engineering practice, one popular system of exhaust heat recovery is called gas turbine-based combined cooling and power (CCP) as shown schematically in Figure 7.4(c). In the plant, the exhaust from the gas turbine transfers heat in an HRST. The steam generated from the HRGS is supplied to a steam turbine-generator to generate electricity and an absorption chiller to produce chilled water.

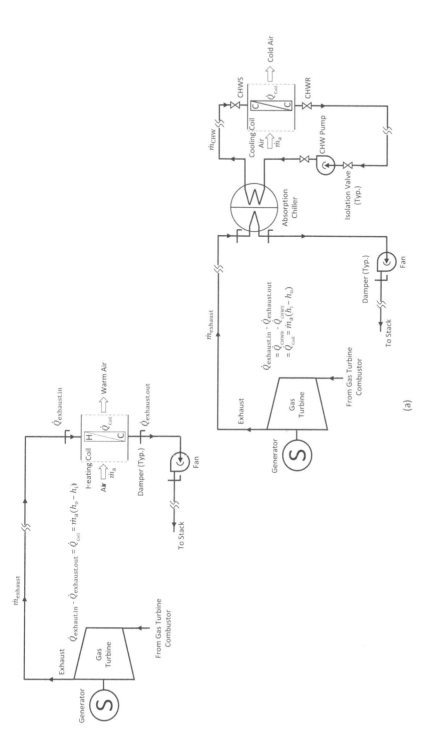

(a)

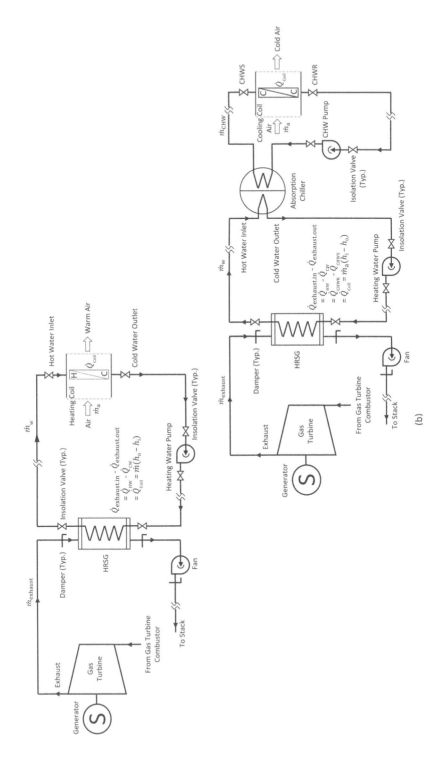

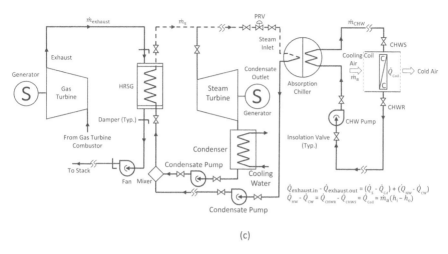

(c)

FIGURE 7.4 Heat recovery from the gas turbine exhaust. (a) Directly to the heat coil or the absorption chiller. (b) An independent hot water loop with a heat exchanger. (c) Combined cooling and power.

7.3.3 STEAM TURBINE EXHAUST

The temperature of exhaust steam from a steam turbine is in the range of 105–120°C. The exhaust steam is condensed to water called condensate in the condenser. The latent heat converting steam to condensate is discharged to the environment by cooling water in a cooling tower. To recover latent heat from the exhaust steam, a recovery system can be used to connect the heating coil directly or the absorption chiller to generate chiller water as shown schematically in Figure 7.5. In Figure 7.5(a), the cooling water used as the hot water is directly pumped to the heating coil for heating. In Figure 7.5(b), cooling water is used as heating energy input in the absorption

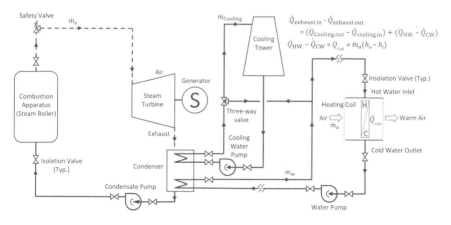

(a)

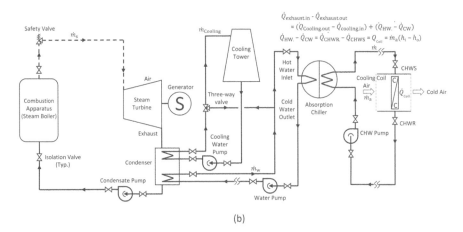

(b)

FIGURE 7.5 Exhaust steam for heating and cooling. (a) The heating system. (b) The cooling system.

chiller to generate chilled water, which circulates through the cooling coil. To stabilize the steam turbine power output, one three-way valve is installed to connect the cooling water pipeline to the cooling tower and the cooling water pipeline as the hot water line for the accommodation of the flows between the two pipelines. Therefore, the impact from the load change of the heating and cooling of air passing through the coil to the turbine power output can be eliminated by manipulating the flow through the three-way valve. When the heating and cooling load of air passing through the coil is lower, more cooling water flows in the cooling water pipeline to the cooling tower, and vice versa.

In engineering applications, the plant that generates power and recovers exhaust steam heat used for the process of heating or cooling is called a cogeneration plant.

7.4 RENEWABLE ENERGIES

7.4.1 SOLAR ENERGY

Solar energy is one of the most abundant and free energy sources on earth. This energy can be applied for electricity generation, heating, and cooling. Solar electricity generation typically uses two technologies: photovoltaics (PV), which generates electricity directly from sunrays, and concentrated solar power (CSP), which uses solar heat to generate electricity in the conventional thermal power plant. Solar heating and cooling (SHC) technologies collect thermal energy from the sun and apply the heat for heating and cooling in residential, commercial, and industrial applications.

7.4.1.1 Photovoltaics

Solar PV technology uses a solar panel, also called a solar cell, to create an electrical current. A major component in the panel is a silicon module of an N type and a P type layer. When sunrays shine on the panel, photons from the rays carry their

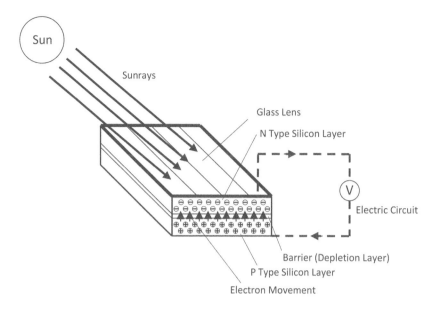

FIGURE 7.6   The photovoltaic solar panel and working principle.

energy down through the cell and give up their energy to electrons in the lower, P type layer. The electrons are triggered to jump across the barrier into the upper, N type layer. Thus, an electric field is built across the layers. By connecting the N type and P type layers with a metallic wire, the electrons travel from the N type layer to the P type layer through the external wire to create an electrical current. Figure 7.6 shows a schematic of the photovoltaic solar panel and working principle.

7.4.1.2   Concentrated Solar Power

CSP technology uses solar heat to generate electricity in a conventional thermal power plant.

A CSP plant requires more equipment and components to generate electricity compared to a PV power plant. Figure 7.7 shows schematics of a direct solar-steam turbine power plant and the steam turbine power plant with a heat transfer fluid (HTF) heat exchanger. In a direct solar-steam power plant, steam generated from the solar collector goes directly to a turbine-generator to generate electricity as shown schematically in Figure 7.7(a). In a CSP plant with an HTF heat exchanger, the HTF absorbs solar heat in the collector and transfers the heat to water flowing through the HTF heat exchanger. Steam is generated in the high-efficiency HTF heat exchanger. The HTF loop and water-steam loop are two independent systems as shown schematically in Figure 7.7(b). To continue providing steam after the sun sets or is covered, solar energy storage provides a workable solution. There are several strategies of solar energy storage—Figure 7.7(c) shows a schematic of a two-tank solar heat storage system. In operation, extra heat from a solar collector is transferred from the HTF to the storage fluid in the hot fluid storage tank. After sunset or when overcast, the hot storage fluid discharges heat back to the HTF.

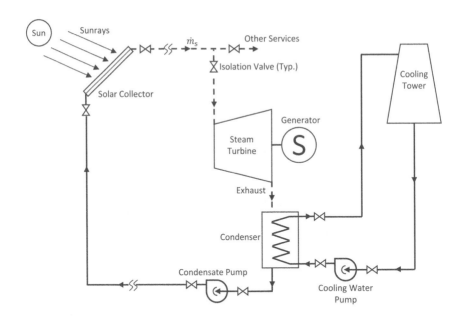

(a)

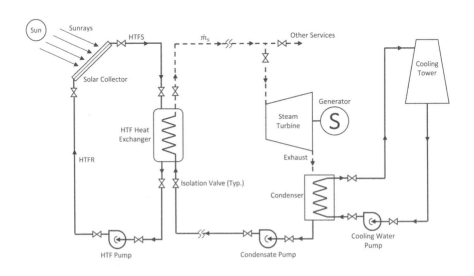

(b)

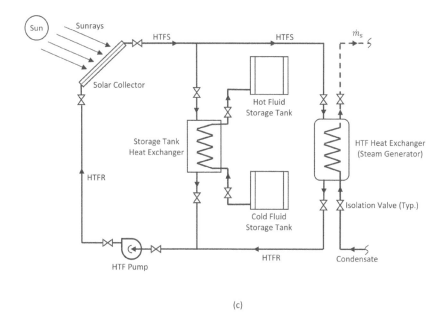

(c)

FIGURE 7.7   CSP plants. (a) A direct solar-steam turbine power plant. (b) A plant with an HTF heat exchanger. (c) A plant with an HTF heat exchanger and heat storage tanks.

The solar energy collectors in CSP plants are concentrated types. There are a variety of concentrated solar collectors, such as parabolic trough, linear Fresnel, parabolic dish, and solar power tower as shown schematically in Figure 7.8. Generally, concentrated solar collectors are equipped with solar tracking systems.

7.4.1.2.1 Parabolic Trough

Figure 7.8(a) shows a schematic of the parabolic trough collector. The collector uses aligned polished-metal mirrors (reflectors) with parabola-shaped surfaces to receive incident sunrays and reflect them onto a common focal absorber tube (receiver) above

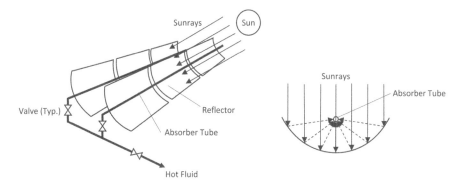

(a)

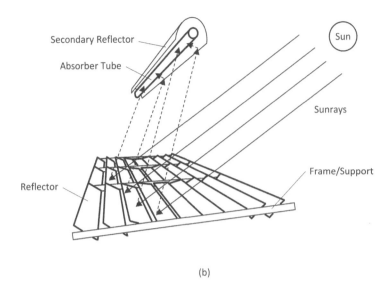

(b)

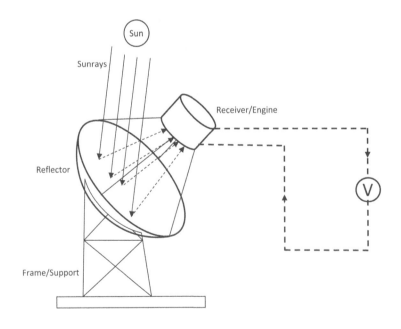

(c)

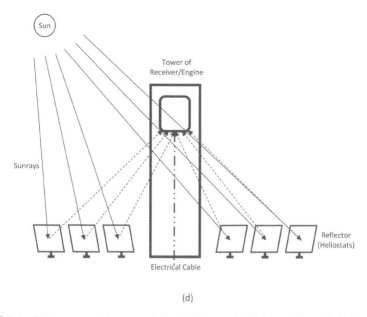

(d)

FIGURE 7.8 Solar power collectors. (a) Parabolic trough. (b) Linear Fresnel. (c) Parabolic dish. (d) Solar power tower.

the mirrors. The heat from sunrays reflected from mirrors is transferred to fluid flowing inside the tube. The fluid temperature can reach quite high when the fluid leaves the tube. The rate of solar heat from the reflector to the receiver is expressed as,

$$\dot{Q}_r = \eta_{opt} A_a G \tag{7.2}$$

where η_{opt} is the optical efficiency of the mirror, A_a is the surface area of the mirror, and G is the irradiation of solar rays in W/m². The rate of heat loss from the collector to the surroundings is determined to be

$$\dot{Q}_{loss} = U A_r \left(T_c - T_a \right) \tag{7.3}$$

where U is the overall heat transfer coefficient of the receiver in W/m², A_r is the surface area of the receiver, and T_c and T_a are the temperatures of the collector and surrounding air, respectively. Useful heat transferred to the fluid, therefore, is

$$\dot{Q}_{useful} = \dot{Q}_r - \dot{Q}_{loss} = \eta_{opt} A_a G - U A_r (T_c - T_a)$$

The efficiency of the solar collector is defined as the ratio of useful heat delivered to the fluid to the radiation incident on the collector,

$$
\begin{aligned}
\eta_c &= \frac{\dot{Q}_{useful}}{\dot{Q}_{incident}} = \frac{\eta_{opt} A_a G - U A_r \left(T_c - T_a \right)}{A_a G} \\
&= \eta_{opt} - \frac{U A_r \left(T_c - T_a \right)}{A_a G} = \eta_{opt} - \frac{U \left(T_c - T_a \right)}{CR \times G}
\end{aligned} \tag{7.4}
$$

where CR is called the concentrator factor which is the ratio of A_a/A_r. The larger of η_{opt} and CR, the higher the solar collector efficiency.

7.4.1.2.2   Linear Fresnel

Figure 7.8(b) shows a schematic of the linear Fresnel collector. The collector uses the aligned long-flat segment of mirrors to receive incident sunrays and reflect them onto a common focal absorber tube (receiver) above the mirrors. The working principle of the linear Fresnel collector is the same as that of the parabolic trough collector.

7.4.1.2.3   Parabolic Dish

Figure 7.8(c) shows a schematic of the parabolic dish collector. The collector has a mirror surface, which is a circular paraboloid, to receive incident sunrays and reflect them onto a common focal absorber tube (receiver) above the mirror. The working principle of the parabolic dish collector is the same as that of the parabolic trough collector. The fluid at high temperature exiting the receiver goes to a Stirling engine-generator mounted on the receiver.

7.4.1.2.4   Solar Power Tower

Figure 7.8(d) shows a schematic of the solar power tower. The tower is a tall structure housing a focal receiver on the top of the tower. The receiver collects the sunrays reflected from an array of aligned flat mirrors known as heliostats on the ground. Fluid flowing in the receiver tube absorbs heat from the sunrays and transfers heat to the water in a separate water-steam loop through a high-efficiency heat exchanger. The steam generated in the heat exchanger eventually goes to the steam turbine-generator on the tower or on the ground to generate electricity.

Example 7.3

A solar power plant has a total reflector (heliostats) surface area of 85,000 m² that collects solar irradiation of 980 W/m² to produce a steam rate of 25 kg/s at 3,000 kPa and 425°C. The steam is supplied to a steam turbine-generator to generate electricity. If the pressure of the exhaust steam from the turbine is 25 kPa, the isentropic efficiency of the steam turbine is 86%, and the power consumption from the other equipment in the plant is neglected, determine (a) the turbine power output (kW), (b) the thermal efficiency of the plant based on the solar irradiation, and (c) the annual cost savings ($/yr) of the equivalent fuel knowing that the low heating value of the fuel oil is 42,000 kJ/kg, the fuel unit cost is $1.51, and the plant operates 3,500 hrs/yr.

SOLUTION

The states of the steam entering and leaving the steam turbine are specified by the given temperature and pressure. Referring to Appendix A.5 Properties of Steam and Compressed Water, the following properties of the steam entering and leaving the steam turbine are found,

$$h_1 = 3,288.1 \; \frac{kJ}{kg}$$

$$s_1 = 7.0015 \; \frac{kJ}{kg.K}$$

and

$s_2 = s_1$ (isentropic process) at 25 kPa

$$x = \frac{s_1 - s_f}{s_g - s_f} = \frac{(7.0015 - 0.8926) \dfrac{kJ}{kg.K}}{(7.8310 - 0.8926) \dfrac{kJ}{kg.K}} = 0.881$$

$$h_{s2} = x h_{fg} - h_f$$

$$= (0.881 \times 2347.4) \frac{kJ}{kg} + 267.97 \frac{kJ}{kg} = 2,336.0 \frac{kJ}{kg}$$

From the isentropic efficiency equation,

$$\eta_s = (h_1 - h_2)/(h_1 - h_{2s})$$

$$h_2 = h_1 - \eta_s (h_1 - h_{2s})$$

$$= 3,288.1 \frac{kJ}{kg} - 0.86 (3,288.1 - 2,336.0) \frac{kJ}{kg} = 2,469.3 \frac{kJ}{kg}$$

(a) The turbine power output is

$$\dot{W}_{out} = \dot{m}(h_1 - h_2) = \left(25 \frac{kg}{s}\right)(3,288.1 - 2,469.3) \frac{kJ}{kg}$$

$$= \mathbf{20,470\ kW}$$

(b) The thermal efficiency of the plant is

$$\eta_{th} = \frac{\dot{W}_{out}}{AG} = \frac{20,470\ kW}{85,000\ m^2 \times 0.98 \dfrac{kW}{m^2}}$$

$$= 0.2457 = \mathbf{24.57\%}$$

(c) The equivalent fuel consumption is determined to be

$$AG = LHV\dot{G}$$

$$\dot{G} = \frac{AG}{LHV} = \frac{(85,000\ m^2)\left(0.98 \dfrac{kW}{m^2}\right)}{42,000 \dfrac{kJ}{kg}}$$

$$= 1.9833 \frac{kg}{s} = 119 \frac{kg}{min}$$

Therefore, the annual cost savings of the equivalent fuel is

$$\left(1.51 \frac{\$}{kg}\right)\left(119 \frac{kg}{min}\right)\left(3,500 \frac{hrs}{yr}\right)\left(60 \frac{min}{hr}\right)$$

$$= 37.74 \times 10^6 \frac{\$}{yr} = \mathbf{37.74} \frac{\mathbf{million\ \$}}{\mathbf{yr}}$$

7.4.1.3 Heating and Cooling

Solar heating and cooling are attractive and popular. An important component in solar heating and cooling is the solar panel. The panel is a flat-plate solar collector that mainly consists of copper tubes and heat-absorbing material inside an insulated frame. The frame is covered with a glass called glazing on the top to receive the solar irradiation. A typical solar panel is shown in Figure 7.9(a). The heat-absorbing material has a special coating on the surface for absorbing solar heat effectively. The glass traps heat in the panel that otherwise radiates out, which is similar to how a greenhouse works. The work principle of the solar panel is that sunrays travel through the glass and strike the surface of the heat-absorbing materials and cold fluid flowing inside the tubes absorbs the heat to become hot fluid. Hot fluid is pumped to the heating coil for heating or to the absorption chiller as heating energy input for cooling. Figures 7.9(b) and (c) show schematics of the solar heating and cooling systems, respectively. In engineering practice, solar heating and cooling can be combined in

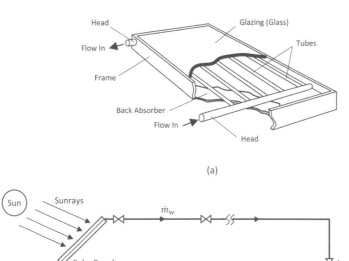

(a)

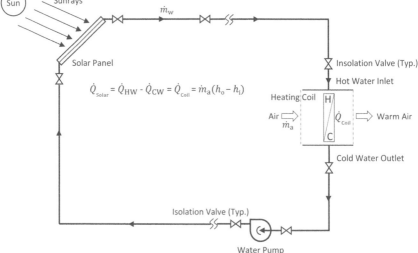

$$\dot{Q}_{Solar} = \dot{Q}_{HW} - \dot{Q}_{CW} = \dot{Q}_{Coil} = \dot{m}_a(h_o - h_i)$$

(b)

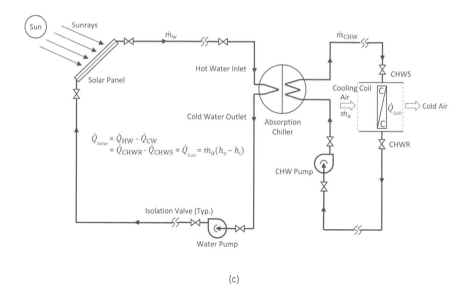

(c)

(d)

FIGURE 7.9 Solar heating and cooling systems. (a) A solar heating panel. (b) Solar heating. (c) Solar cooling. (d) A combined solar heating and cooling system.

one system. Figure 7.9(d) shows a schematic of a combined solar heating and cooling system.

The rate of solar heat absorbed by the absorber plate is

$$\dot{Q}_{\text{abs}} = \tau \alpha A G \tag{7.5}$$

where A is the area of the collector surface in m² and $\tau \alpha$ is a product of transmissivity-absorptivity. Heat loss from the collector to the surrounding is

$$\dot{Q}_{\text{loss}} = UA(T_c - T_a) \tag{7.6}$$

TABLE 7.1

Properties of Solar Heating Collectors

	$\tau\alpha$	U (W/m².°C)
No glazing	0.90	28
Single glazing	0.85	2.8
Double glazing	0.80	1.7

Table 7.1 shows typical values of transmissivity-absorptivity product $\tau\alpha$ and the overall heat transfer efficiency U of the solar heating collectors.

The useful heat transferred to the fluid in the panel, therefore, is

$$
\begin{aligned}
\dot{Q}_{useful} &= \dot{Q}_{abs} - \dot{Q}_{loss} = \tau\alpha AG - UA(T_c - T_a) \\
&= A\left[\tau\alpha G - U(T_c - T_a)\right]
\end{aligned}
\tag{7.7}
$$

In terms of the fluid flowing in the solar panel, the useful heat is equal to

$$
\dot{Q}_{useful} = \dot{m}_f c_p\left(T_{f,out} - T_{f,in}\right) = \dot{m}_f\left(h_{f,out} - h_{f,in}\right)
$$

where $\dot{m}_f$ is the fluid flow rate, and $T_{f,in}$ and $T_{f,out}$ are the temperatures of fluid entering and exiting the panel, respectively. The efficiency of a solar panel is determined to be

$$
\begin{aligned}
\eta_c &= \frac{\dot{Q}_{useful}}{\dot{Q}_{incident}} = \left[\tau\alpha AG - UA(T_c - T_a)\right]/AG \\
&= \tau\alpha - U(T_c - T_a)/G
\end{aligned}
\tag{7.8}
$$

Example 7.4

A solar heating panel with a double glazing connected to an air heating coil has an efficiency of 48%. Measurement shows the solar incident is 300 W/m² and the ambient temperature is 23°C. If the temperature of air entering the coil is 7°C T_{db}, 50% RH at a flowrate of 70 m³/min, and the temperature of air leaving the coil is 22°C T_{db}, determine (a) the required panel surface area and (b) the average temperature of the collector.

SOLUTION

The states of air entering and leaving the coil are specified by the given temperatures and relative humidity. Referring to Appendix A.4 Psychrometric Chart at aa Pressure of 1 atm (101.325 kPa), the following properties of the air entering and leaving the coil are found,

$$
h_1 = 15.01\ \frac{kJ}{kg}
$$

$$
v_1 = 0.7965\ \frac{m^3}{kg}
$$

and

$$h_2 = 31.45 \frac{kJ}{kg}$$

The air mass flowrate is

$$\dot{m} = \frac{\dot{V}_1}{v_1} = \frac{70 \frac{m^3}{min}}{0.7965 \frac{m^3}{kg}} = 87.88 \frac{kg}{min}$$

Heat transferred to the air is

$$\dot{Q} = \dot{m}(h_2 - h_1) = \left(87.88 \frac{kg}{min}\right)\left(31.45 \frac{kJ}{kg} - 15.01 \frac{kJ}{kg}\right)$$

$$= 1,444.75 \frac{kJ}{min}$$

Without considering heat loss in the fluid system, there is a heat balance,

$$\dot{Q} = \dot{Q}_{useful}$$

(a) From the efficiency equation of the solar panel, the required surface area of the solar collector can be decided as

$$A = \frac{\dot{Q}_{useful}}{\eta_c G} = \frac{1,444.75 \frac{kJ}{min}}{0.48\left(0.3 \frac{kW}{m^2}\right)}$$

$$= 167.2 \text{ m}^2$$

(b) From Equation (7.7), the average temperature of the panel, therefore, is determined to be

$$T_c = \frac{\tau \alpha G - \dot{Q}_{useful}/A}{U} + T_a$$

$$= \frac{0.8\left(0.3 \frac{kW}{m^2}\right) - 1,444.75 \frac{kJ}{min}/167.3 \text{ m}^2}{1.7 \frac{W}{m^2 °C}} + 23°C$$

$$= 79.51°C$$

7.4.1.4 Combined Cooling, Heating, and Power Generation

In engineering practice, cooling, heating, and power generation can be operated in one plant. Such a plant is called a combined cooling, heating, and power (CCHP) plant. The CCHP plant has a higher overall thermal efficiency than the total efficiency of the plants separately for power, cooling, and heating. There are a variety of solar-powered CCHP plant configurations. Figure 7.10 schematically shows some of them:

1. Steam generated from the solar collector directly goes to a CCHP plant as shown in Figure 7.10(a).

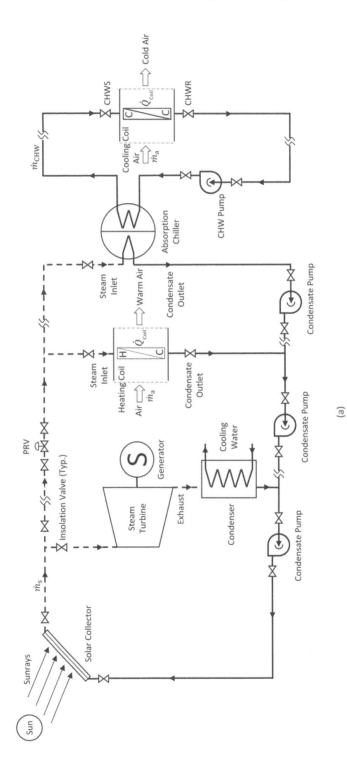

(a)

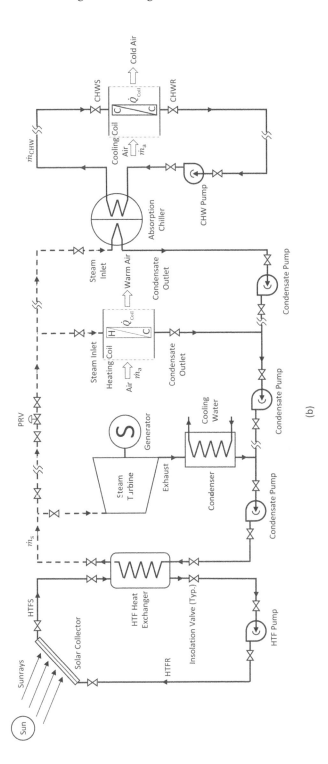

(b)

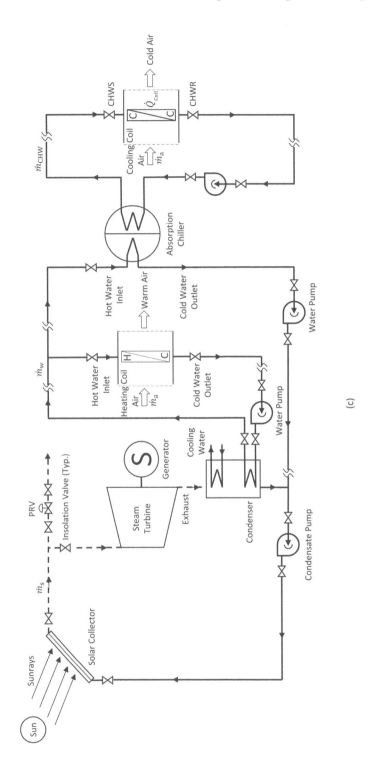

(c)

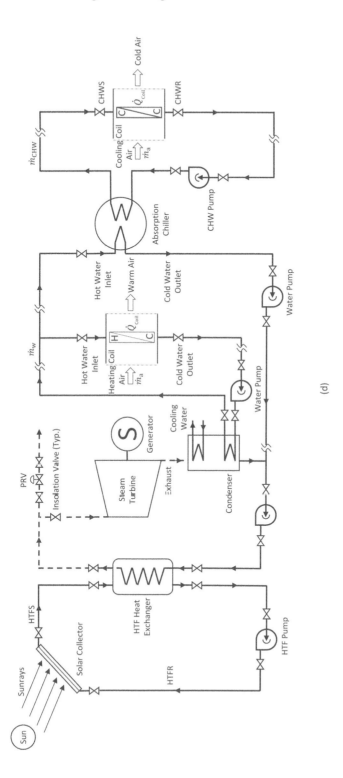

(d)

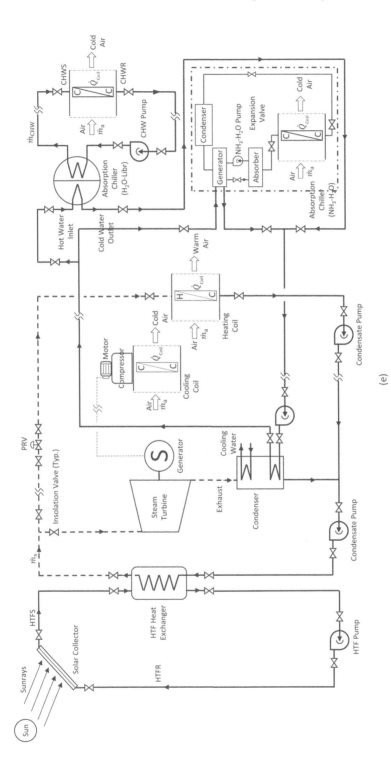

FIGURE 7.10 Solar-powered CCHP plants. (a) A steam CCHP. (b) A CCHP with an HTF heat exchanger generating steam for heating and cooling. (c) A CCHP using hot water heated by the exhaust steam for heating and cooling. (d) A CCHP with an HTF heat exchanger and using hot water for heating and cooling. (e) A CCHP with an HTF heat exchanger and using hot water to charge absorption chillers.

2. Steam generated from an HTF heat exchanger goes to a CCHP plant as shown in Figure 7.10(b).
3. A CCHP plant using heat recovered from turbine exhaust steam for heating and cooling as shown in Figure 7.10(c).
4. A CCHP plant using hot water generated from the turbine exhaust steam for heating and cooling with an HTF heat exchanger as shown in Figure 7.10(d).
5. A CCHP plant using hot water generated from the turbine exhaust steam to charge absorption chillers (H_2O-LiBr and NH_3-H_2O) with an HTF heat exchanger as shown in Figure 7.10(e).

7.4.2 GEOTHERMAL ENERGY

Geothermal energy is thermal energy existing underground. The energy is the heat from the earth and hot fluid deep in the earth. Figure 7.11 shows the temperature distribution in the earth depth.

7.4.2.1 Heating and Cooling

A simple application of geothermal energy for heating and cooling is to lay a pipeline underground. Fluid circulates inside the pipeline to absorb heat from the earth for heating or to release heat to the earth for cooling. Figure 7.12(a) shows schematics of three types of pipelines: horizontal pipeline, vertical pipeline, and ground-water well. Figure 7.12(b) shows heating and cooling modes of the geothermal heat pump system, respectively.

Hot geothermal fluid can also be pumped out from the deep earth. In such an application, a heat exchanger may be applied to transfer geothermal energy to the fluid in a separate loop. Figure 7.13(a) shows schematics of geothermal heating and cooling by using hot geothermal fluid. Figure 7.13(b) shows a schematic of the combined geothermal heating and cooling by using hot geothermal fluid.

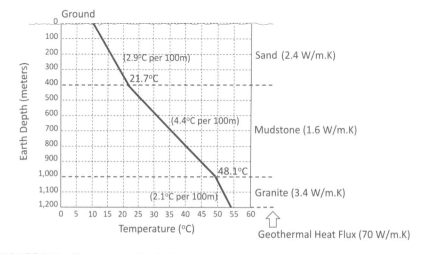

FIGURE 7.11   Temperature distribution in the earth depth.

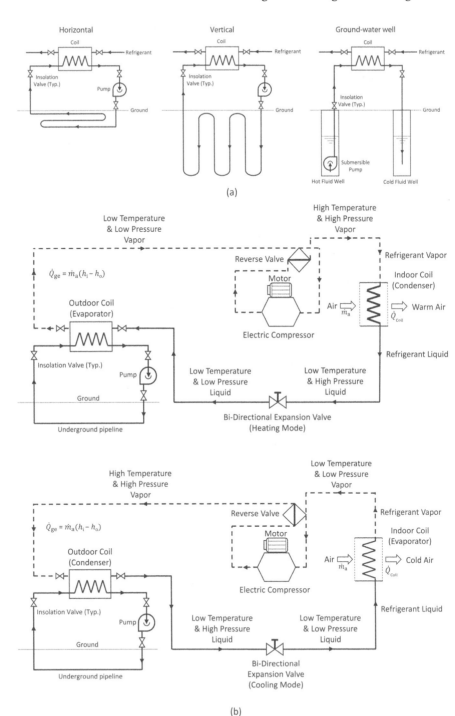

FIGURE 7.12 Underground pipelines and modes of geothermal heat pump system. (a) Underground pipeline layout. (b) Heating and cooling mode: Heating mode (top) and cooling mode (bottom).

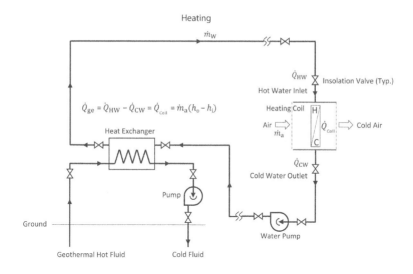

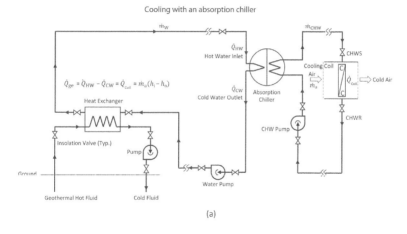

(a)

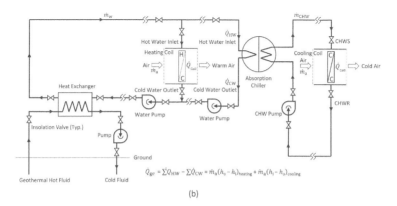

(b)

FIGURE 7.13   Heating, cooling and combined heating and cooling. (a) Heating and cooling.
(b) Combined heating and cooling.

7.4.2.2   Power Generation

Figure 7.14 shows schematics of the thermal power plants that harvest energy from geothermal hot fluid to generate electricity. There are four types: single-flash plant, double-flash plant, binary cycle plant, and combined flash-binary plant.

7.4.2.2.1   Single-Flash Plant

In the single-flash cycle plant as shown in Figure 7.14(a), the hot geothermal fluid at a mass flow rate of m_f is pumped out of the earth and sent to a flash tank at a specified pressure. The two-phase mixture produced in the flash tank is then separated into liquid and steam in a separator. The steam is directed to a steam turbine-generator to generate electricity. The exhaust steam in the turbine is condensed to condensate in a condenser. Finally, the m_w of the condensate mixed with the geothermal liquid from the separator is discharged back to the earth.

7.4.2.2.2   Double-Flash Plant

In the double-flash cycle plant as shown in Figure 7.14(b), the hot geothermal fluid at a mass flow rate of m_f is pumped out of the earth. After a flash tank, the steam produced in a separator is directed to a steam turbine-generator to generate electricity. The geothermal liquid leaving the separator is further expanded in a second flash tank. Additional steam resulting from the second flash tank is directed to a lower pressure stage of the turbine. The rest of the processes after the steam turbine are the same as those in the single-flash plant.

7.4.2.2.3   Binary Cycle Plant

In the binary cycle plant as shown in Figure 7.1(c), the geothermal fluid and turbine working medium flow in two separate systems. The working medium works in a closed system. The hot geothermal fluid transfers heat to the working medium in a heat exchanger. The steam generated in the heat exchange goes to a steam turbine-generator to generate electricity. The geothermal fluid after the heat exchanger is discharged back to the earth. Exhaust steam in the turbine is condensed to condensate in a condenser. Then, condensate is pumped back to the heat exchanger to repeat the cycle.

7.4.2.2.4   Combined Flash-Binary Plant

The combined flash-binary plant as shown in Figure 7.14(d) is developed by incorporating the binary cycle in the single-flash plant. Hot geothermal fluid is separated into steam and liquid in a separator. The steam is directed to a main steam turbine-generator and the liquid flows to a heat exchanger in which the liquid transfers heat to the working medium in an independent closed system. The steam produced in the closed system is directed to a secondary steam turbine-generator. Finally, the m_w of the condensate from a main condenser mixed with the geothermal liquid from the heat exchanger is discharged back to the earth.

In engineering practices, heating and cooling can be combined with the power generation in one geothermal plant as shown schematically in Figure 7.15. Figure 7.15(a) shows a schematic of the cogeneration plant having steam heating and cooling combined with a double-flash geothermal power generation. Figure 7.15(b) shows

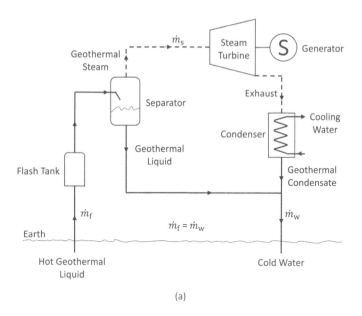

(a)

(b)

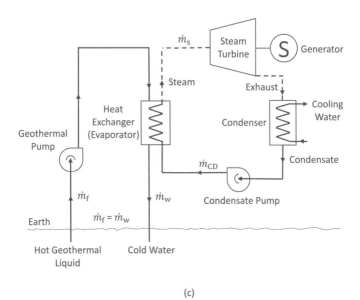

(c)

(d)

FIGURE 7.14 Geothermal power plants. (a) The single-flash cycle. (b) The double-flash cycle. (c) The binary cycle. (d) The combined flash-binary plant.

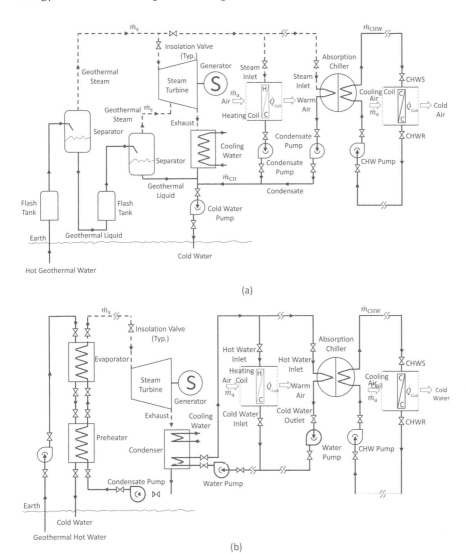

FIGURE 7.15 Geothermal cogeneration plants. (a) Heating and cooling combined with a double-flash cycle. (b) Heating and cooling combined with a binary cycle.

a schematic of the cogeneration plant having hot water heating and cooling with a binary cycle geothermal power generation.

7.4.3 OCEAN THERMAL ENERGY

Ocean thermal energy is heat from the ocean's water temperature. In tropical and equatorial regions, surface water temperature can reach up to 25°C and deep water temperature can be as low as 4°C. Figure 7.16 shows the water temperature distribution with the ocean depth.

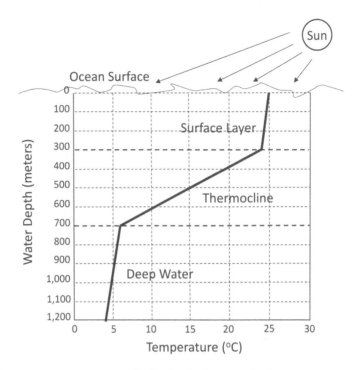

FIGURE 7.16   Water temperature distribution in the ocean depth.

7.4.3.1   Heating and Cooling

Figure 7.17 shows a schematic of heating and cooling with a heat pump utilizing ocean thermal energy. During cold weather, the refrigerant liquid circulated in the heat pump absorbs heat from warm surface water in the evaporator and becomes vapor. The refrigerant vapor is at high pressure after the compressor and transfers the heat to air passing through the indoor coil in the condenser. In hot weather, the refrigerant liquid circulated in the heat pump absorbs heat in the evaporator from air passing through the indoor coil and becomes vapor. The refrigerant vapor is at high pressure after the compressor and discharges heat to cold deep water through the outdoor coil in the condenser.

7.4.3.2   Power Generation

The technology that harnesses ocean thermal energy to generate electricity is called ocean thermal energy conversion (OTEC). OTEC uses the temperature difference between warm surface water and cold deep water to operate the power cycle. Two basic power cycles of OTEC—closed-cycle and open-cycle—are shown in Figures 7.18(a) and (b). Two modified cycles based on the basic cycles—hybrid-cycle and two-stage cycle—are shown in Figures 7.18(c) and (d).

7.4.3.2.1   Closed-Cycle

Closed-cycle OTEC uses warm surface water to heat and vaporize a refrigerant (working fluid) at a low-boiling point, such as NH_3 (ammonia) or R-134a, in a heat

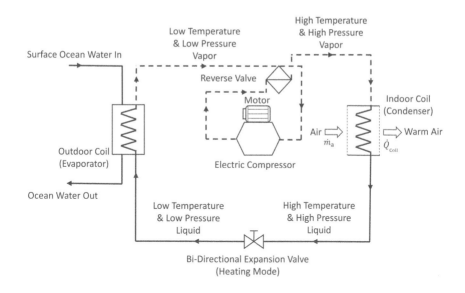

(a)

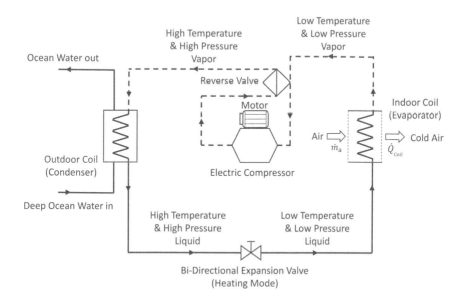

(b)

FIGURE 7.17   Heating and cooling with a heat pump. (a) Heating. (b) Cooling.

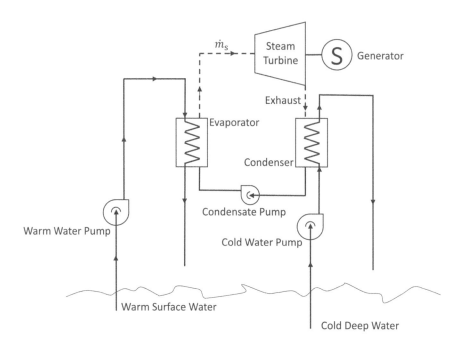

(a)

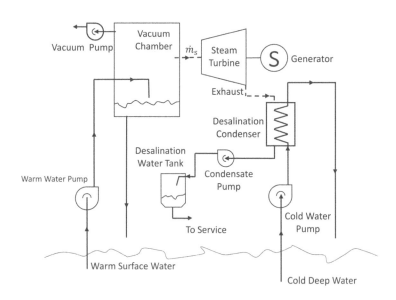

(b)

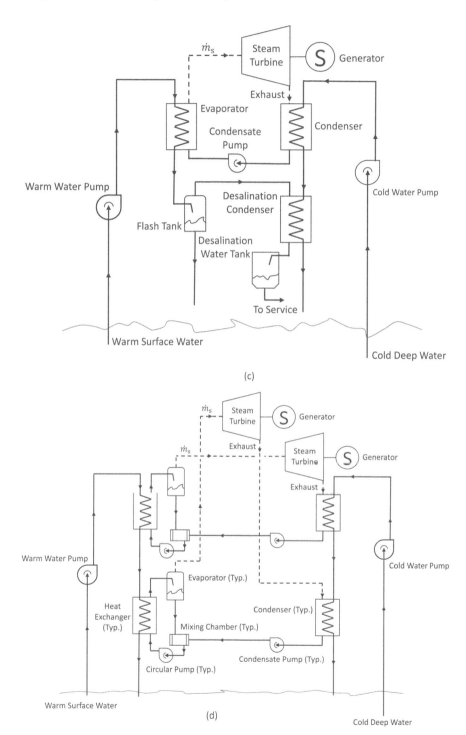

FIGURE 7.18 Cycles of OTEC. (a) The closed-cycle. (b) The hybrid-cycle. (c) The hybrid-cycle. (d) The two-stage cycle.

exchanger (evaporator). The vapor then expands in a steam turbine. The turbine in turn drives a generator to produce electricity. The vapor becomes condensate in a condenser by using cold deep water as cooling water. The condensate is discharged out of the condenser and pumped back to the heat exchanger to repeat the cycle.

7.4.3.2.2 Open-Cycle

Open-cycle OTEC uses steam from the warm surface water to directly generate electric power. The warm surface water is first pumped into a vacuum chamber to be flashed to steam. Then, the steam drives a steam turbine-generator to produce electricity. Steam is condensed to desalinated water in a condenser by using the cold deep water as cooling water. The desalinated water is then collected in a tank. Open-system OTEC plant can not only produce electricity, but can also provide desalinated water for drinking, irrigation, aquaculture, etc.

7.4.3.2.3 Hybrid-Cycle

Hybrid-cycle OTEC combines the features of the closed- and open cycle. The warm surface water heats and vaporizes a refrigerant (working fluid) at a low-boiling point, such as NH_3 (ammonia) or R-134a, in a heat exchanger (evaporator). The refrigerant vapor drives a steam turbine-generator to produce electricity, while the ocean surface water enters a flash tank after the heat exchanger to generate vapor. The flashed vapor is cooled by the cold deep water in a desalination condenser to become desalinated water. The desalinated water is then collected in a tank to be supplied for drinking, irrigation, aquaculture, etc.

7.4.3.2.4 Two-Stage Cycle

In a two-stage cycle OTEC, two independent closed cycles are aligned serially with the warm surface water flow. Both cycles consist of identical equipment. The warm surface water passes through two heat exchangers in series, respectively. Heat is transferred from the warm surface water to the refrigerant (working fluid) at a low-boiling point, such as NH_3 (ammonia) or R-134a, in heat exchangers, one after another. Similarly, cold deep water passes through two heat exchangers (condensers) in series to cool the vapor exhausted from the turbines to be condensated. In the closed cycle, the refrigerant is heated in the heat exchanger and separated into vapor and liquid in the evaporator (flash tank). The vapor then flows toward the turbine and drives the turbine-generator to generate electricity. Condensate out of the condenser is discharged to the mixing chamber and pumped back to the heat exchanger with refrigerant from the evaporator to repeat the cycles.

7.4.3.3 Plant Location

An OTEC plant can have an option to be built in three locations: on land, offshore, or on a floating ship. Figure 7.19 shows schematics of the three plant locations. Each option has its own advantages and disadvantages.

7.4.3.3.1 On Land
- Less concern regarding mooring and plant stability.
- Less work of maintenance and repair farther out in water.

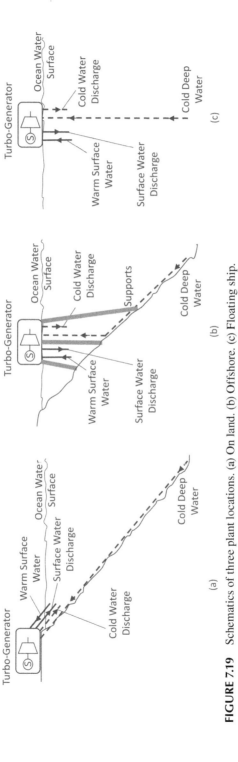

FIGURE 7.19 Schematics of three plant locations. (a) On land. (b) Offshore. (c) Floating ship.

- Plant byproduct, i.e, desalinated water, is conveniently supplied to home, irrigation, or aquaculture.
- A long, cold, deep water pipe may be needed.

7.4.3.3.2 Offshore
- Less concern regarding mooring and lengthy pipes.
- More work to build a platform and support structure.
- Power delivery to land becomes costly due to requiring long underwater cables to land.

7.4.3.3.3 Floating Ship
- Flexible location.
- Difficult mooring and stabilizing the plant.
- Power delivery to land becomes costly due to requiring long underwater cables to land.

Figure 7.20 shows a schematic of an electric network from an offshore OTEC plant to land users. In the figure, the user is an air conditioner with a cooling coil that is driven by electricity generated from the offshore OTEC plant.

The maximum thermal efficiency of the OTEC turbine-generator with a 21°C (temperatures of warm surface water and cold deep water are 25°C and 4°C, respectively) temperature difference is

$$\eta_{th} = \frac{\text{Plant Net Power Output}}{\text{Total Required Energy Input}}$$

$$= 1 - \frac{Q_L}{Q_H} = 1 - \frac{T_L}{T_H} = 1 - \frac{(4 + 273)\,\text{K}}{(25 + 273)\,\text{K}}$$

$$= 1 - \frac{277\text{K}}{298\text{K}} = 0.0705 = 7.05\%$$

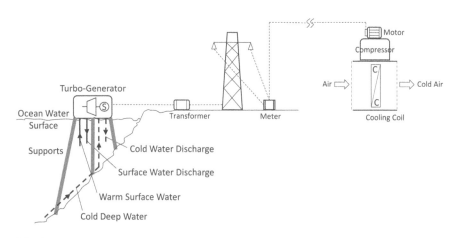

FIGURE 7.20   Electric network from an offshore OTEC plant.

The overall efficiency of the OTEC plant is actually quite smaller than 7.05% since a certain amount of the electricity generated from the turbine-generator needs to be consumed internally in the plant, such as by driving pumps and powering auxiliary equipment.

7.4.4 Wind Energy

Wind energy is kinetic energy from moving air. A wind turbine is used to convert the kinetic energy from wind to generate electricity. Ideally, the maximum wind energy obtained from a wind stream blowing through a wind turbine is:

$$\dot{E} = \frac{1}{2}\dot{m}V^2 = \frac{1}{2}(\rho AV)V^2 = \frac{1}{2}\rho AV^3 \tag{7.9}$$

where $\dot{m}$ is the wind mass flowrate, ρ is the density of air, A is the turbine swept area, and V is the wind velocity entering the turbine. Many wind turbines can work together in one area to connect to a common electric power network. A group of many wind turbines in the area is called a wind farm. A wind farm may consist of a dozen to hundreds of wind turbines. There are many varieties of turbine shapes available. In general, they are in two categories: horizontal-axis wind turbines (HAWT) and vertical-axis wind turbines (VAWT).

7.4.4.1 Horizontal-Axis and Vertical-Axis Wind Turbines

The difference between a HAWT and a VAWT is the turbine shaft axis to the wind flow direction. In the HAWT, the turbine shaft axis is parallel with the wind flow direction. In other words, the shaft axis of the HAWT is parallel with the ground. While in the VAWT, the turbine shaft axis is perpendicular to the wind flow direction. In other words, the shaft axis of the VAWT is vertical to the ground. Figure 7.21 shows schematics of the HAWT and VAWT, respectively.

7.4.4.2 Wind Power System

The wind turbine is a wind power plant. Wind brings kinetic energy to the wind turbine. The turbine drives a generator to produce electrical power. Figure 7.21 shows

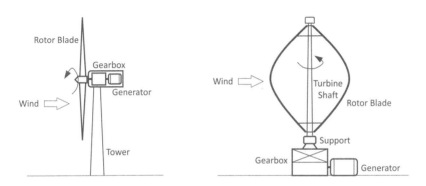

FIGURE 7.21   HAWT and VAWT.

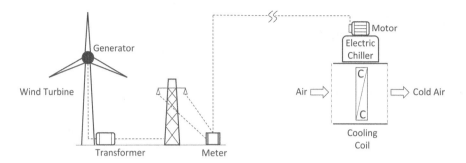

FIGURE 7.22   Electric network from an HAWT power plant.

a schematic of an air conditioner with a cooling coil powered by electricity supplied from an HAWT power plant.

Example 7.5

A wind power turbine with a rotor blade radius of 5 m is located at a field where the wind blowing velocity is 5 m/s at 1 atm and 40°C. If the electrical power output of the turbine is 810 kW and the efficiency of the generator is 92%, determine (a) the maximum wind energy rate (kW) and (b) the wind turbine efficiency.

SOLUTION

Referring to Appendix A.3 Specific Heat of Air, the gas constant of air R is 0.287 kJ/kg.K and referring to Appendix A.1 Unit Conversion, 0.287 kJ/kg.K is equal to 0.287 $\frac{kPa.m^3}{kg.K}$. Using Equation (1.1) (see Chapter 1), the density of the air is

$$\rho = \frac{P}{RT} = \frac{101.325 \text{ kPa}}{\left(0.287 \dfrac{kPa.m^3}{kg.K}\right)(40 + 273)\,°C} = 1.128 \frac{kg}{m^3}$$

The swept area of the wind turbine is

$$A = \pi r^2 = 3.1416(5 \text{ m})^2 = 78.54 \text{ m}^2$$

(a) Using Equation (7.9), the maximum wind energy rate is

$$\dot{E} = \frac{1}{2}\rho A V^3 = \frac{1}{2}\left(1.128 \frac{kg}{m^3}\right)(78.54 \text{ m}^2)\left(5 \frac{m}{s}\right)^3$$

$$= \left(5,537.07 \frac{kg.m^2}{s^3}\right)\left[\frac{1 \text{ N}}{1 \text{ kg.}\left(\dfrac{m}{s^2}\right)}\right] = \left(5,537.07 \frac{N.m}{s}\right)\left(\frac{1 \text{ J}}{1 \text{ N.m}}\right)$$

$$= 5.54 \frac{kJ}{s} = \textbf{5.54 kW}$$

(b) The turbine shaft work is determined to be

$$\dot{W}_{shaft} = \frac{\dot{W}_{electic}}{\eta_{generator}} = 810 \text{ kW} / 0.92 = 880.43 \text{ kW}$$

The wind turbine efficiency, therefore, is determined to be

$$\eta_{turbine} = \frac{\dot{W}_{shaft}}{\dot{E}} = \frac{880.43 \text{ kW}}{5,537.07 \text{ kW}} = 0.159 = \textbf{15.9\%}$$

Appendix

A.1   Unit Conversion

Dimension	Metric	Metric/English
Acceleration	$1 \text{ m/s}^2 = 100 \text{ cm/s}^2$	$1 \text{ m/s}^2 = 3.2808 \text{ ft/s}^2$
		$1 \text{ ft/s}^2 = 0.3048 \text{ m/s}^2$
Area	$1 \text{ m}^2 = 10^4 \text{ cm}^2 = 10^6 \text{ mm}^2 = 10^{-6} \text{ km}^2$	$1 \text{ m}^2 = 1550 \text{ in}^2 = 10.764 \text{ ft}^2$
		$1 \text{ ft}^2 = 144 \text{ in}^2 = 0.09290304 \text{ m}^2$
Density	$1 \text{ g/cm}^3 = 1 \text{ kg/L} = 1000 \text{ kg/m}^3$	$1 \text{ g/cm}^3 = 62.428 \text{ lbm/ft}^3 = 0.036127 \text{ lbm/in}^3$
		$1 \text{ lbm/in}^3 = 1728 \text{ lbm/ft}^3$
		$1 \text{ kg/in}^3 = 0.062428 \text{ lbm/ft}^3$
Energy, heat, work, internal energy, enthalpy	$1 \text{ kJ} = 1000 \text{ J} = 1000 \text{ N.m} = 1 \text{ kPa.m}^3$	$1 \text{ kJ} = 0.94782 \text{ Btu}$
	$1 \text{ kJ/kg} = 1000 \text{ m}^2/\text{s}^2$	$1 \text{ Btu} = 1.055056 \text{ kJ} = 5.40395 \text{ paia.ft}^3$
	$1 \text{ kWh} = 3600 \text{ kJ}$	$= 778.169 \text{ lbf.ft}$
	$1 \text{ cal} = 4.1868 \text{ J}$	$1 \text{ Btu/lbm} = 25037 \text{ ft}^2/\text{s}^2 = 2.326 \text{ kjkJ/kg}$
	$1 \text{ IT cal} = 4.1868 \text{ J}$	$1 \text{ kJ/kg} = 0.430 \text{ Btu/lbm}$
	$1 \text{ Cal} = 4.1868 \text{ J}$	$1 \text{ kWh} = 3412.14 \text{ Gtu}$
		$1 \text{ therm} = 10^5 \text{ Btu} = 1.055 \times 10^5 \text{ kJ}$ (natural gas)
Force	$1 \text{ N} = 1 \text{ kg.m/s}^2 = 10^5 \text{ dyne}$	$1 \text{ N} = 0.22481 \text{ lbf}$
	$1 \text{ kgf} - 9.80665 \text{ N}$	$1 \text{ lbf} = 32.174 \text{ lbm.ft/s}^? = 4.44822 \text{ N}$
Heat flux	$1 \text{ W/cm}^2 = 10^4 \text{ W/m}^2$	$1 \text{ W/m}^2 = 0.3171 \text{ Btu/h.ft}^2$
Heat transfer coefficient	$1 \text{ W/m}^2.°C = 1 \text{ W/m}^2.\text{K}$	$1 \text{ W/m}^2.°C = 0.17612 \text{ Btu/h.ft}^2. °F$
Length	$1 \text{ m} = 100 \text{ cm} = 1000 \text{ mm} = 10^6 \text{ μm}$	$1 \text{ m} = 39.270 \text{ in} = 3.2808 \text{ ft} = 1.0926 \text{ yd}$
	$1 \text{ km} = 1000 \text{ m}$	$1 \text{ ft} = 12 \text{ in} = 0.3048 \text{ m}$
		$1 \text{ mile} = 5280 \text{ ft} = 1.6093 \text{ km}$
		$1 \text{ in} = 2.54 \text{ cm} = 25.4 \text{ mm}$
Mass	$1 \text{ kg} = 1000 \text{ g}$	$1 \text{ kg} = 2.2046226 \text{ lbm}$
	$1 \text{ metric ton} = 1000 \text{ kg}$	$1 \text{ bm} = 0.45359237 \text{ kg}$
		$1 \text{ ounce} = 28.3495 \text{ g}$
		$1 \text{ slug} = 32.174 \text{ lbm} = 14.5939 \text{ kg}$
		$1 \text{ short ton} = 2000 \text{ lbm} = 907.1847 \text{ kg}$
Power, heat transfer rate	$1 \text{ W} = 1 \text{ J/s}$	$1 \text{ kW} = 3412.14 \text{ Btu/h} = 737.56 \text{ lbf.ft/s}$
	$1 \text{ kW} = 1000 \text{ W} = 1.341 \text{ hp}$	$1 \text{ hp} = 550 \text{ ibf.ft/s} = 0.7068 \text{ Btu/s} = 42.41 \text{ Btu/min} = 2544.5 \text{ Btu/h} = 0.7457 \text{ kW}$
	$1 \text{ hp} = 745.7 \text{ W}$	$1 \text{ boiler hp} = 33475 \text{ Btu/h}$

(Continued)

Dimension	Metric	Metric/English
		1 Btu/h = 1.055056 kJ/h
		1 ton of refrigeration = 200 Btu/min
Pressure, pressure head	1 Pa = 1 N/m^2	1 Pa = 1.4504 × 10^{-4} psia = 0.020886 lbf/ft^2
	1 kPa = 10^3 Pa = 10^{-3} MPa	1 psi = 144 lbf/ft^2 = 6.894757 kPa
	1 atm = 101.325 kPa = 1.01325 bars = 760 mm Hg@0°C = 1.03323 kgf/cm^2	1 atm = 14.696 psia = 29.92 in Hg@30°F 1 in Hg = 3.387 kPa = 13.60 in H$_2$O
	1 mm Hg = 0.1333kPa	
Specific heat	1 kJ/kg.°C = 1 kJ/kg.K = 1 J/g.°C	1 Btu/lbm.°F = 4.1868 kJ/gk.°C
		1 Btu/lbmol.R = 4.1868 kJ/kmol.K
		1 kJ/kg.°C = 0.23885 Btu/lbm.°F = 0.23885 Btu/lbm.R
Specific volume	1 m^3/kg = 1000 L/kg = 1000 cm^3/g	1 m^3/kg = 16.02 ft^3/lbm
		1ft^3/lbm = 0.062428 m^3/kg
Temperature	T(K) = T(°C) + 273.15	T(R) = T(°F) + 459.67 = 1.8 × T(K)
	ΔT(K) = ΔT(°C)	T(°F) = 1.8 × T(°C) + 32
		ΔT(°F) = ΔT(R) = 1.8 × ΔT(K)
Thermal conductivity	1 W/m.°C = 1 W/m.K	1 W/m.°C = 0.57782 Btu/h.ft. °F
Thermal resistance	1°C/W = 1 kW	1 kW = 0.52750°F.h/Btu
Velocity	1 m/s = 3.60 km/h	1 m/s = 3.2808 ft/s = 2.237 mi/h
		1 mi/h = 1.46667 ft/s
		1 mi/h = 1.6093 km/h
Viscosity, dynamic	1 kg/m.s = 1 N.s/m2 = 1 Pa.s = 10 poise	1 kg/m.s = 2419.1 lbm/ft.h = 0.020886 lbf.s/ft^2 = 0.67197 lbm/fr.s
Viscosity, kinematic	1 m^2/s = 10^4 cm^2/s	1 m^2/s = 10.764 ft^2/s = 3.875 × 10^4 ft^2/h
	1 stoke = 1 cm^2/s = 10^{-4} m^2/s	
Volume	1 m^3 = 1000 L = 10^6 cm^3 (cc)	1 m^3 = 6.1024 × 10^4 in^3 = 10^6 cm^3 = 35.315 ft^3 = 264.17 gal (U.S.)
		1 gal (U.S.) = 231 in^3 = 3.7854 L
		1 gal (U.S.) = 128 fl ounce
		1 fl ounce = 29.5735 cm^3 = 0.0295735 L
Volume flow rate	1 m^3/s = 60000 L/min = 10^6 cm^3/s	1 m^3/s = 15850 gal/min (gpm) = 35.315 ft^3/s = 2118.9 ft^3/min (CFM)

A.2 Some Physical Constants and Commonly Used Properties

Physical Constant	Metric	English
Universal gas constant	R_u = 8.31447 kJ/kmol.K = 8.31447 kPa.m^3/kmol.K = 0.0831447 bar. m^3/kmol.K = 82.05 L.atm/kmol.K	R_u = 1.9858 Btu/lbmol.R = 1545.35 ft.lbf/lbmol.R = 10.73 psia.ft^3/lbmol.R
Standard acceleration of gravity	g = 32.174 ft/s^2	g = 32.174 ft/s^2
Standard atmospheric pressure	1 atm = 101.325 kPa = 1.01325 bar = 760 mm Hg@0°C = 10.3323 m H$_2$O@4°C	1 atm = 101.325 kPa = 1.01325 bar = 14.696 psia = 760 mm Hg@0°C = 29.9213 in Hg@32°F 10.3323 m H$_2$O@4°C
Stefan-Boltzmann constant	σ = 5.670 × 10^{-8} W/m^2.K^4	σ = 0.1714 × 10^{-8} Btu/h.ft^2.R^4
Enthalpy of vaporization of water @ 1 atm	h_{fg} = 2256.5 kJ/kg	h_{fg} = 970.12 Btu/lbm

Properties	Metric	English
Air @ 20°C (68°F) and 1 atm		
Specific gas constant	R_{air} = 0.2870 kJ/kg.K = 287.0 m^2/s^2.K	R_{air} = 0.06855 Btu/lbm.R = 53.34 ft.lbf/bm.R = 1716 ft^2/s^2.R
Specific heat ratio	$k = c_p/c_v$ = 1.40	$k = c_p/c_v$ = 1.41
Speed of sound	c = 343.2 m/s = 1236 km/h	c = 1126 ft/s = 767.7 mi/h
Density	ρ = 1.204 kg/m^3	ρ = 0.07518 lbm/ft^3
Dynamic viscosity	μ = 1.825 × 10^{-5} kg/m.s	μ = 1.227 × 10^{-5} lbm/ft.s
Kinematic viscosity	v = 1.516 × 10^{-5} m^2/s	v = 1.632 × 10^{-4} ft^2/s
Liquid water @ 20°C (68°F) and 1 atm		
Specific heat $(c = c_p = c_v)$	c = 4.182 kJ/kg.K = 4182 m^2/s^2.K	c_p = 0.9989 Btu/lbm.R = 777.3 ft.lbf/lbm.R = 25009 ft^2/s^2.R
Density	ρ = 998.0 kg/m^3	ρ = 62.3 lbm/ft^3
Dynamic viscosity	μ = 1.002 × 10^{-3} kg/m.s	μ = 6.733 × 10^{-4} lbm/ft.s
Kinematic viscosity	v = 1.004 × 10^{-6} m^2/s	v = 1.081 × 10^{-5} ft^2/s

A.3   Specific Heat of Air

Temperature T (K)	Gas Constant R (kJ/kg.K)	Density ρ (kg/m³)	Specific Heat at Constant Pressure c_p (kJ/kg.K)	Specific Heat at Constant Volume c_v (kJ/kg.K)	Specific Hear Ratio k
240	0.2870	1.4710	1.006	0.7164	1.404
260		1.3579	1.006	0.7168	1.403
273.2		1.2923	1.006	0.7171	1.403
280		1.2609	1.006	0.7173	1.402
288.7		1.2229	1.006	0.7175	1.402
300		1.1768	1.006	0.7180	1.402
320		1.1033	1.007	07192	1.400
340		1.0384	1.009	0.7206	1.400
360		0.9807	1.010	0.7223	1.398
380		0.9291	1.012	0.7243	1.397
400		0.8826	1.014	0.7266	1.396

A.4 Psychrometric Chart at a Pressure of 1 atm (101.325 kPa)

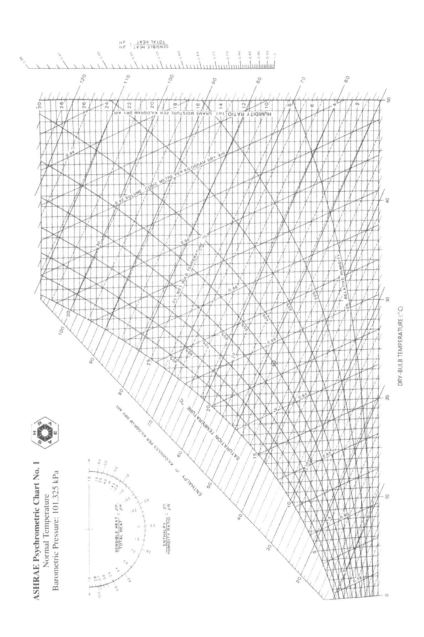

ASHRAE Psychrometric Chart No. 1
Normal Temperature
Barometric Pressure: 101.325 kPa

A.5 Properties of Steam and Compressed Water

Saturated Steam (Temperature)

		Density, kg/m³		Enthalpy, kJ/kg			Entropy, kJ/(kg·K)			Volume, cm³/g	
T, °C	P, MPa	ρ_L	ρ_V	h_L	h_V	Δh	S_L	S_V	Δ_L	v_L	v_V
20	0.0023393	998.16	0.017314	83.91	2537.4	2453.5	0.29648	8.6660	8.3695	1.00184	57757
25	0.0031699	997.00	0.023075	104.83	2546.5	2441.7	036711	8.5566	8.1894	1.00301	43337
30	0.0042470	995.61	0.30415	125.73	2555.5	2429.8	0.43675	8.4520	8.0152	1.00441	32878
35	0.0056290	993.99	0.39674	146.63	2564.5	2417.9	0.50513	8.3517	7.8466	1.00605	25205
40	0.0073849	992.18	0.051242	167.53	2573.5	2406.0	0.57240	8.2555	7.6831	1.00789	19515
45	0.0095950	990.17	0.065565	188.43	2582.4	2394.0	0.63861	8.1633	7.5247	1.00992	15252
50	0.012352	988.00	0.083147	209.34	2591.3	2381.9	0.70381	8.0748	7.3710	1.01215	12027
55	0.015762	985.66	0.10456	230.26	2600.1	2369.8	0.76802	7.9898	7.2218	1.01455	9564.3
60	0.019946	983.16	0.13043	251.18	2608.8	2357.7	0.83129	7.9081	7.0769	1.01713	7667.2
65	0.025042	980.52	0.16146	272.12	2617.5	2345.4	0.89365	7.8296	6.9359	1.01987	6193.5
70	0.031201	977.73	0.19843	293.07	2626.1	2333.0	0.95513	7.7540	6.7989	1.02277	5039.5
75	0.038595	974.81	0.24219	314.03	2634.6	2320.6	1.0158	7.6812	6.6654	1.02584	4128.9
80	0.047414	971.77	0.29367	335.01	2643.0	2308.0	1.0756	7.6111	6.5355	1.02905	3405.2
85	0.057867	968.59	0.35388	356.01	2651.3	2295.3	1.1346	7.5434	6.4088	1.03243	2825.8
90	0.070182	965.30	0.42390	377.04	2659.5	2282.5	1.1929	7.4781	6.2853	1.03595	2359.1
95	0.084608	961.88	0.50491	398.09	2667.6	2269.5	1.2504	7.4151	6.1647	1.03963	1980.6
100	0.10142	958.35	0.59817	419.17	2675.6	2256.4	1.3072	7.3541	6.0469	1.04346	1671.8
105	0.12090	954.70	0.70503	440.27	2683.4	2243.1	1.3633	7.2952	5.9318	1.04744	1418.4
110	0.14338	950.95	0.82693	461.42	2691.1	2229.6	1.4188	7.2381	5.8193	1.05158	1209.3
115	0.16918	947.08	0.96540	482.59	2698.6	2216.0	1.4737	7.1828	5.7091	1.05588	1035.8
120	0.19867	943.11	1.1221	503.81	2705.9	2202.1	1.5279	7.1391	5.6012	1.06033	891.21
125	0.23224	939.02	1.2987	525.07	2713.1	2188.0	1.5816	7.0770	5.4955	1.64.94	770.03
130	0.27028	934.83	1.4970	546.38	2720.1	2173.7	1.6346	7.0264	5.3918	1.69.71	668.00
135	0.31323	930.54	1.7190	567.74	2726.9	2159.1	1.6872	6.9772	5.2900	1.07465	581.73
140	0.36154	926.13	1.9667	589.16	2733.4	2144.3	1.7392	6.9293	5.1901	1.07976	508.45
145	0.41568	921.62	2.2423	610.64	2739.8	2129.2	1.7907	6.8826	5.0919	1.8504	445.96
150	0.47616	917.01	2.5481	632.18	2745.9	2113.7	1.8418	6.8371	4.9953	1.99050	392.45
155	0.54350	912.28	2.8863	653.79	2751.8	2098.0	1.8924	6.7926	4.9002	1.09615	346.46
160	0.61823	907.45	3.2596	675.47	2757.4	2082.0	1.9426	6.7491	4.8066	1.10199	306.78
165	0.70093	902.51	3.6707	697.24	2762.8	2065.6	1.9923	6.7066	4.7143	1.10803	272.59
170	0.79219	897.45	4.1222	719.08	2767.9	2048.8	2.0417	6.6650	4.6233	1.12072	216.58
175	0.89260	892.28	4.6172	741.02	2772.7	2031.7	2.0906	6.6241	4.5335	1.12740	173.90
180	1.0028	887.00	5.1588	763.05	2777.2	2014.2	2.1392	6.5840	4.4448	1.12740	193.84
185	1.1235	881.60	5704	785.19	2781.4	1996.2	2.1875	6.5447	4.3571	1.13430	173.90
190	1.2552	876.08	6.3954	807.43	2785.3	1977.9	2.2355	6.5059	4.2704	1.1445	156.36
195	1.3988	870.43	7.0976	829.79	2788.8	1959.0	2.2832	6.4678	4.1846	1.14886	140.89
200	1.5549	864.66	7.8610	852.27	2792.0	1939.7	2.3305	6.4302	4.0996	1.15653	127.21
205	1.7243	858.76	8.6898	874.88	2794.8	1919.9	2.3777	6.3930	4.0154	1.16448	115.08
210	1.9077	852.72	9.5885	897.63	2797.3	1899.6	2.4245	6.3563	3.9318	1.17272	104.29
215	2.1058	846.54	10.562	920.53	2799.3	1878.8	2.4712	6.3200	3.8488	1.18125	94.679
220	2.3196	840.22	11.615	943.58	2800.9	1857.4	2.5177	6.2840	3.7663	1.19017	86.092

(Continued)

Saturated Steam (Temperature) (*Continued*)

T, °C	P, MPa	Density, kg/m³		Enthalpy, kJ/kg			Entropy, kJ/(kg·K)			Volume, cm³/g	
		ρ_L	ρ_V	h_L	h_V	Δh	S_L	S_V	Δ_L	v_L	v_V
225	2.5497	833.75	12.755	966.80	2802.1	1835.4	2.5640	6.2483	3.6843	1.19940	78.403
230	2.7971	827.12	13.985	990.19	2802.9	1812.7	2.6101	6.2128	3.6027	1.20902	71.503
235	3.0625	820.33	15.314	1013.8	2803.2	1789.4	2.6561	6.1775	3.5214	1.21902	65.298
240	3.3469	813.37	16.749	1037.6	2803.0	1765.4	2.7020	6.1423	3.4403	1.22946	59.705
245	3.6512	806.22	18.297	1061.5	2802.2	1740.7	2.7478	6.1072	3.3594	1.24035	54.654
250	3.9762	798.89	19.967	1085.8	2800.9	1715.2	2.7935	6.0721	3.2785	1.25173	50.083
255	4.3229	791.37	21.768	1110.2	2799.1	1688.8	2.8392	6.0369	3.1977	1.26364	45.938
260	4.6923	783.63	23.712	1135.0	2796.6	1661.6	2.8849	6.0016	3.1167	1.27612	42.173
265	5.0853	775.66	25.809	1160.0	2793.5	1633.5	2.9307	5.9661	3.0354	1.28922	38.746
270	5.5030	767.46	28.073	1185.3	2789.7	1604.4	2.9765	5.9304	2.9539	1.30300	35.621
275	5.9464	759.00	30.520	1210.9	2785.2	1574.3	3.0224	5.8944	2.8720	1.31751	32.766
280	6.4166	750.00	33.165	1236.9	2779.9	1543.0	3.0685	5.8579	2.7894	1.33284	30.153
285	6.9147	741.25	36.028	1263.2	2773.7	1510.5	3.1147	5.8209	2.7062	1.34907	27.756
290	7.4418	731.91	39.132	1290.0	2766.7	1476.7	3.1612	5.7834	2.6222	1.36630	25.555
295	7.9991	722.21	42.501	1317.3	2758.7	1441.4	3.2080	5.7451	2.5371	1.38464	23.529
300	8.5879	712.14	46.168	1345.0	2749.6	1404.6	3.2552	5.7059	2.4507	1.40423	21.660
305	9.2094	701.64	50.167	1373.3	2739.4	1366.1	3.3028	5.6657	2.3629	1.42524	19.933
310	9.8651	690.67	54.541	1402.0	2727.9	1325.7	3.3510	5.6244	2.2734	1.44787	18.335
315	10.556	679.18	59.344	1431.8	2715.1	1283.2	3.3998	5.5816	2.1818	1.47236	16.851
320	11.284	667.09	64.638	1462.2	2700.6	1238.4	3.4494	5.5372	2.0878	1.49904	15.471
325	12.051	654.33	70.506	1493.5	2684.3	1190.8	3.5000	5.4908	1.9908	1.52829	14.183
330	12.858	640.77	77.050	1525.9	2666.0	1140.2	3.5518	5.4422	1.8903	1.56061	12.979
335	13.707	626.29	84.407	1559.5	2645.4	1085.9	3.6050	5.3906	1.7856	1.59671	11.847
340	14.601	610.67	92.759	1594.5	2621.8	1027.3	3.6601	5.3356	1.6755	1.63755	10.781
341	14.785	607.38	94.570	1601.8	2616.8	1015.0	3.6714	5.3241	1.6527	1.64640	10574
342	14.971	604.04	96.433	1609.1	2611.5	1002.5	3.6828	5.3124	1.6296	1.65551	10.370
343	15.159	600.64	98.351	1616.4	2606.1	989.7	3.6943	5.3005	1.6063	1.66490	10.168
344	15.349	597.17	100.33	1623.9	2600.6	976.7	3.7059	5.2885	1.5826	1.67457	9.9674
345	15.541	593.63	102.36	1631.5	2594.9	963.4	3.7176	5.2762	1.5586	1.68456	9.7690
346	15.734	590.01	104.47	1639.1	2589.0	949.9	3.7295	5.2636	1.5342	1.69488	9.5724
347	15.930	586.32	106.64	1646.9	2583.0	936.1	3.7414	5.2509	1.5094	1.70556	9.3776
348	16.128	582.54	108.88	1654.8	2576.7	922.0	3.7536	5.2379	1.4843	1.71661	9.1844
349	16.328	578.67	111.20	1662.8	2570.3	907.5	3.7659	5.2246	1.4587	1.72810	8.9927
350	16.529	574.71	113.61	1670.9	2563.6	892.7	3.7784	5.2110	1.4326	1.74002	8.8024
351	16.733	570.64	116.10	1679.1	2556.8	877.6	3.7910	5.1971	1.4061	1.75243	8.6134
352	16.939	566.46	118.68	1687.5	2549.6	862.1	3.8039	5.1829	1.3790	1.76536	8.4257
353	17.147	562.15	121.37	1696.1	2542.3	846.2	3.8170	5.1683	1.3514	1.77888	8.2390
354	17.358	557.72	124.17	1704.8	2534.6	829.8	3.8303	5.1534	1.3231	1.79302	8.0533
355	17.570	553.14	127.09	1713.7	2526.6	812.9	3.8439	5.1380	1.2942	1.80786	7.8684
356	17.785	548.41	130.14	1722.8	2518.4	795.5	3.8577	5.122	1.2645	1.82347	7.6841
357	18.002	543.50	133.33	1732.2	2509.8	777.6	3.8719	5.1059	1.2340	1.83993	7.5003
358	18.221	538.41	136.67	1741.7	2500.8	759.0	3.8864	5.0891	1.2026	1.85733	7.3168
359	18.442	533.11	140.19	1751.5	2491.4	739.8	3.9014	5.0717	1.1703	1.87578	7.1332
360	18.666	527.59	143.90	1761.7	2481.5	719.8	3.9167	5.0536	1.1369	1.89541	6.9493

(Continued)

Saturated Steam (Temperature) (*Continued*)

T, °C	P, MPa	Density, kg/m³		Enthalpy, kJ/kg			Entropy, kJ/(kg·K)			Volume, cm³/g	
		ρ_L	ρ_V	h_L	h_V	Δh	S_L	S_V	Δ_L	v_L	v_V
361	18.892	521.82	147.82	1772.1	2471.1	699.0	3.9325	5.0347	1.1023	1.91635	6.7649
362	19.121	515.79	151.99	1782.9.	2460.2	677.3	3.9488	5.0151	1.0663	1.93879	6.5795
363	19.352	509.45	156.43	1794.1	2448.6	654.5	3.9656	4.9945	1.0288	1.96290	6.3925
364	19.582	502.78	161.20	1805.7	2436.2	630.5	3.9831	4.9727	0.9896	1.98894	6.2035
365	19.821	495.74	166.35	1817.8	2422.9	605.2	4.0014	4.9497	0.9483	2.0172	6.0115
366	20.060	488.27	171.95	1830.5	2408.7	578.2	4.0205	4.9251	0.9046	2.0480	5.8157
367	20.302	488.27	178.11	1843.8	2393.1	549.2	4.0406	4.8986	0.8580	2.0821	5.6145
368	20.546	471.67	184.98	1858.1	2375.9	517.8	4.0621	4.8697	0.8076	2.1201	5.4061
369	20.793	462.18	192.77	1873.5	2356.6	483.1	4.0853	4.8376	07523	2.1636	5.1875
370	21.044	451.43	201.84	1890.7	2234.5	443.8	4.1112	4.8012	0.6901	2.2152	4.9544
371	21.297	438.64	212.79	1910.6	2308.3	397.7	4.1412	4.7586	0.6175	2.2798	4.6995
372	21.554	422.26	226.84	1935.3	2275.5	340.3	4.1785	4.7059	0.5274	2.3682	4.4084
373	21.814	398.68	247.22	1969.7	2229.8	260.1	4.2308	4.6334	0.4026	2.5083	4.0450
t_e	22.064	322.00	322.00	2084.3	2084.3	0.	4.4070	4.4070	0.	3.1056	3.1056

Saturated Steam (Pressure)

$p,$ MPa	$T,$ °C	Density, kg/m³		Enthalpy, kJ/kg			Entropy, kJ/(kg·K)			Volume, cm³/g	
		ρ_L	ρ_V	h_L	h_V	Δh	S_L	s_V	Δ_s	v_L	v_V
0.0010	6.970	999.86	0.007741	29.30	2513.7	2484.4	0.10591	8.9749	8.8690	1.00014	129178
0.0014	11.969	999.46	0.10650	50.28	2522.8	2472.5	0.18015	2.8521	8.6719	1.00054	93899
0.0018	15.837	998.93	0.012511	66.49	2529.9	2463.4	0.23662	8.7608	8.5241	1.00108	74011
0.0024	20.414	998.08	0.17738	85.49	2538.2	2452.5	0.30239	8.6567	8.3544	1.00193	56375
0.0032	25.158	996.96	0.023282	105.49	2546.8	2441.3	0.36945	8.5533	8.1838	1.00305	42952
0.0040	28.960	995.92	0.028743	121.39	2553.7	2432.3	0.42239	8.4734	8.0510	1.00110	34791
0.0050	32.874	994.70	0.035480	137.75	2560.7	2415.2	0.47620	8.3938	7.9176	1.00533	28185
0.0060	36.159	993.59	0.42135	151.48	2566.6	2408.4	0.52082	8.3290	7.8082	1.00645	23733
0.0070	39.000	992.55	0.048722	163.35	2571.7	2408.4	0.55903	8.2745	7.7154	1.00750	20524
0.0080	41.509	991.59	0.055252	173.84	2576.2	2402.4	0.59249	8.2273	7.6348	1.00848	18099
0.0090	43.761	990.69	0.061731	183.25	2580.2	2397.0	0.62230	8.1858	7.5635	1.00940	16199
0.010	45.806	989.83	0.068166	191.81	2583.9	2392.1	0.64920	8.1488	7.0752	1.01027	14670
0.020	60.058	983.13	0.13075	251.42	2608.9	2357.5	0.83202	7.9072	6.6429	1.01716	7648.0
0.040	75.587	974.30	0.25044	317.62	2636.1	2318.4	1.0261	7.6690	6.5018	1.02638	3993.0
0.050	81.317	970.94	0.25044	340.54	2645.2	2304.7	1.0912	7.5930	6.5018	1.02993	3240.0
0.060	85.926	967.99	0.36607	359.91	2652.9	2292.9	1.1454	7.5311	6.3857	1.03307	2731.7
0.070	89.932	965.34	0.42287	376.75	2659.4	2282.7	1.1921	7.4790	6.2869	1.03590	2364.8
0.080	93.486	962.93	0.47914	391.71	2665.2	2273.5	1.2330	7.4339	6.2009	1.03850	2087.1
0.10	99.606	958.63	0.59034	417.50	2674.9	2257.4	1.3028	7.3588	6.0561	1.04315	1693.9
0.15	111.349	949.92	0.86260	467.13	2693.1	2226.0	1.4337	7.2230	5.7893	1.05273	1159.9
0.20	120.210	942.94	1.1291	504.70	2706.2	2201.5	1.5302	7.1269	5.5967	1.06052	885.68
0.25	127.411	937.02	1.3915	535.34	2716.5	2181.1	1.6072	7.0524	5.4452	1.06722	718.66
0.30	133.522	931.82	1.6508	561.43	2724.9	2163.5	1.6717	6.9916	5.3199	1.07317	605.76
0.35	138.857	927.15	1.9077	584.26	2732.0	2147.7	1.7274	6.9401	5.2128	1.07857	524.18
0.40	143.608	922.89	2.1627	604.65	2738.1	2133.4	1.7765	6.8955	5.1190	1.08355	462.38
0.42	145.375	921.28	2.5642	612.25	2740.3	2128.0	1.7946	6.8791	5.0846	1.08544	441.62
0.44	147.076	919.72	2.3655	619.58	2742.4	2122.8	1.8120	6.8636	5.0516	1.08709	422.74
0.46	148.716	918.20	2.4666	626.64	2744.4	2117.7	1.8287	6.8487	5.0199	1.08908	405.42
0.48	150.300	916.73	2.5674	633.47	2746.3	2112.8	1.8448	6.8344	4.9895	1.09084	389.50
0.50	151.831	915.29	2.6680	640.09	2748.1	2108.0	1.8604	6.8207	4.9603	1.09255	37481
0.52	153.314	913.89	2.7685	646.50	2749.9	2103.4	1.8754	6.8075	4.9321	1.09423	361.20
0.54	154.753	912.52	2.8688	652.72	2751.5	2098.0	1.8899	6.7948	4.9049	1.09587	348.58
0.56	156.146	911.18	2.9689	658.77	2753.1	2094.4	1.9040	6.7825	4.8786	1.09748	336.82
0.58	157.506	909.87	3.0689	664.65	2754.7	2090.0	1.9176	6.7707	4.8531	1.09905	325.85
0.60	158.826	908.59	3.1687	670.38	2756.1	2085.6	1.9308	6.7592	4.8284	1.10060	315.58
0.62	160.112	907.34	3.2684	675.96	2757.6	2081.6	1.9437	6.7482	4.8045	1.10212	305.96
0.64	161.365	906.11	3.3680	681.41	2758.9	2077.5	1.9562	6.7374	4.7813	1.103.62	296.91
0.66	162.587	904.91	3.4675	686.73	2760.3	2073.5	1.9684	6.7270	4.7587	1.10509	288.40
0.68	163.781	903.72	3.5668	691.92	2761.5	2069.6	1.9802	6.7169	4.7367	1.10654	280.36
0.70	164.946	902.56	3.6660	697.00	2762.8	2065.8	1.9918	6.7071	4.7153	1.10796	272.77
072	166.086	901.42	3.7652	701.97	2763.9	2062.0	2.0031	6.6975	4.6944	1.10936	265.59
0.74	167.200	900.30	3.8642	706.84	2765.1	2058.2	2.0141	6.6882	4.6741	1.11075	258.79
0.76	168.291	899.19	3.9631	711.61	2766.2	2054.6	2.0248	6.6791	4.6243	1.11211	252.33

(Continued)

Saturated Steam (Pressure) (*Continued*)

p, MPa	T, °C	Density, kg/m³		Enthalpy, kJ/kg			Entropy, kJ/(kg·K)			Volume, cm³/g	
		ρ_L	ρ_v	h_L	h_v	Δh	S_L	s_v	Δ_s	v_L	v_v
0.78	169.360	898.10	4.0620	716.28	2767.3	2051.0	2.0354	6.6703	4.6349	1.11346	246.18
0.80	170.406	897.04	4.1608	720.86	2786.3	2047.4	2.0457	6.6616	4.6160	1.11478	240.34
0.82	171.433	895.98	4.2595	725.36	2769.3	2043.9	2.0557	6.6532	4.5975	1.11609	234.77
0.84	172.440	894.94	4.3581	729.78	2770.3	2040.5	2.0656	6.6449	4.5793	1.11739	229.46
0.86	173.428	893.92	4.4567	734.11	2771.2	2037.1	2.0753	6.6369	4.5616	1.11867	224.38
088	174.398	892.91	4.5552	738.37	2772.1	2033.8	2.0847	6.6290	4.5443	1.11993	219.53
0.90	175.350	891.92	4.6536	742.56	2773.0	2030.5	2.0940	6.6213	4.5272	1.12118	218.89
0.92	176.287	890.93	4.7520	746.68	2773.9	2027.2	2.1032	6.6137	4.5106	1.12242	210.44
0.94	177.207	889.96	4.8503	750.73	2774.7	2024.0	2.1121	6.6063	4.4942	1.12364	206.17
0.96	178.112	889.01	4.9486	754.72	2775.5	2020.8	2.1209	6.5991	4.4782	1.124.85	202.08
0.98	179.002	888.06	5.0468	758.65	2776.3	2017.7	2.1296	6.5920	4.4624	1.126.05	198.14
1.00	179.878	887.13	5.1450	762.52	2777.1	2014.6	2.1381	6.5850	4.4470	1.12723	194.36
1.05	182.009	884.84	5.3903	771.94	2778.9	2007.0	2.1587	6.5681	4.4095	1.13014	185.52
1.10	184.062	882.62	5.6354	781.03	2780.6	1999.6	2.1785	6.5520	4.3735	1.13299	177.45
1.15	186.043	880.46	5.8804	789.82	2782.2	1992.4	2.1976	6.5365	4.3390	1.13577	170.06
1.20	187.957	878.35	6.1251	798.33	2783.7	1985.4	2.2159	6.5217	4.3058	1.13850	163.26
1.25	189.809	876.29	6.3698	806.58	2785.1	1978.6	2.2337	6.5074	4.2737	1.14118	156.99
1.30	191.605	874.28	6.6144	814.60	2786.5	1971.9	2.2508	6.4936	4.2428	1.14380	151.19
1.35	193.347	872.31	6.8589	822.39	2787.7	1965.3	2.2674	6.4803	4.2129	1.14638	145.80
1.40	195.039	870.39	7.1034	829.97	2788.8	1958.9	2.2835	6.4675	4.1839	1.14892	140.78
1.45	196.685	868.50	7.3479	837.35	2789.9	1952.6	2.2992	6.4550	4.1559	1.15141	136.09
1.50	198.287	866.65	7.5924	844.56	2791.0	1946.4	2.3143	6.4430	4.1286	1.15387	131.71
1.55	199.848	864.84	7.8369	851.59	2791.9	1940.3	2.3291	6.4313	4.1022	1.15629	127.60
1.60	201.370	863.05	8.0815	858.46	2792.8	1934.4	2.3435	6.4199	4.0765	1.15868	123.74
1.65	202.856	861.30	8.3261	865.17	2793.7	1928.5	2.3575	6.4089	4.0514	1.16103	120.10
1.70	204.307	859.58	8.5708	871.74	2794.5	1922.7	2.3711	6.3981	4.0270	1.16336	116.67
1.75	205.725	857.89	8.8156	878.17	2795.2	1917.0	2.3845	6.3877	4.0032	1.16565	113.43
1.80	207.112	856.22	9.0606	884.47	2795.9	1911.4	2.3975	6.3775	3.9800	1.16792	110.37
1.85	208.469	854.58	903056	890.65	2796.6	1905.9	2.4102	6.3675	3.9573	1.17016	107.46
1.90	209.798	852.96	9.5508	896.71	2797.2	1900.5	2.4227	6.3578	3.9351	1.17238	104.70
1.95	211.101	851.37	9.7962	902.66	2797.8	1895.1	2.4348	6.3483	3.9135	1.17458	102.08
2.0	212.377	849.80	10.042	908.50	2798.3	1889.8	2.4468	6.3390	3.8923	1.17675	99.585
2.1	214.858	846.72	10.533	919.87	2799.3	1879.4	2.4699	6.3210	3.8511	1.18103	94.938
2.2	217.249	843.72	11.026	930.87	2800.1	1869.2	2.4921	6.3038	3.8116	1.18523	90698
2.3	2.19.557	840.79	11.519	941.53	2800.8	1859.3	2.5136	6.2872	3.7736	1.18936	86.815
2.4	221.789	837.92	12.013	951.87	2801.4	1849.6	2.5343	6.2712	3.7369	1.19343	83.244
2.5	223.950	835.12	12.58	961.91	2801.9	1840.0	2.5543	6.2558	3.7015	1.19743	79.949
2.6	226.046	832.37	13.004	971.67	2802.3	18307	2.5736	6.2409	3.6672	1.20138	76.899
2.7	228.080	829.68	13.501	981.18	2802.7	1821.5	2.5924	6.2264	3.6340	120528	74.066
2.8	230.057	827.04	14.000	990.46	2802.9	1812.4	2.6106	6.2124	3.6018	1.20913	71.429
2.9	231.980	824.45	14.500	999.51	2803.1	1803.6	2.6283	6.1988	3.5705	1.21293	68.968
3.0	233.853	821.90	15.001	1008.3	2803.2	1794.8	2.6455	6.1856	3.5400	1.21669	66.664
3.1	235.679	819.39	15.503	1017.0	2803.2	1786.2	2.6623	6.1727	3.5104	1.22042	64.504

(*Continued*)

Saturated Steam (Pressure) (*Continued*)

p, MPa	T, °C	Density, kg/m³		Enthalpy, kJ/kg			Entropy, kJ/(kg·K)			Volume, cm³/g	
		ρ_L	ρ_V	h_L	h_V	Δh	S_L	s_V	Δ_s	v_L	v_V
3.2	237.459	816.92	16.006	1025.4	2803.1	1777.7	2.6787	6.1602	3.4815	1.22410	62.475
3.3	239.198	814.49	16.512	1033.7	2803.0	1769.3	2.6946	6.1479	6.4533	1.22776	60.564
3.4	240.897	812.10	17.018	1041.8	2802.9	1761.0	2.7102	6.1360	3.4258	1.23138	58.761
3.5	242.557	809.74	17.526	1049.8	2802.6	1752.8	2.7254	6.1243	3.3989	1.23497	57.058
3.6	244.182	807.41	18.036	1057.6	2802.4	1744.8	2.7403	6.1129	3.3726	1.23854	55.446
3.7	245.772	805.10	18.547	1065.3	2802.1	1736.8	2.7549	6.1018	3.3469	1.24208	53.918
3.8	247.330	802.83	19.059	1072.8	2801.7	1728.9	2.7691	6.0908	3.3217	1.24559	52.467
3.9	248.857	80059	19.574	1080.2	2801.3	1721.1	2.7831	6.0801	3.2970	1.24908	51.089
4.0	250.354	798.37	20.090	1087.5	2800.8	1713.3	2.7968	6.0696	3.2728	1.25256	49.776
4.1	251.823	796.17	20.608	1094.7	2800.3	1705.7	2.8102	6.0592	3.2491	1.25601	48.525
4.2	253.264	794.00	21.127	1101.7	2799.8	1698.1	2.8234	60491	3.2257	1.25944	47.332
4.3	254.680	791.85	21.649	1108.7	2799.2	1960.6	2.8363	6.0391	3.2028	1.26286	46.192
4.4	256.070	789.73	22.172	1115.5	2798.6	1683.1	2.8490	6.0293	3.1803	1.26626	45.102
4.5	257.437	787.62	22.697	1122.2	2797.9	1675.7	2.8615	6.0197	3.1582	1.26965	44.059
4.6	258.780	785.53	23.224	1128.9	2797.3	1668.4	2.8738	6.0102	3.1364	1.27302	43.059
4.7	260.101	783.47	23.753	1135.5	2796.5	1661.1	2.8859	6.0009	3.1150	1.27638	42.100
4.8	261.402	781.42	24.284	1141.9	2795.8	1653.9	2.8978	5.9917	3.0939	1.27973	41.180
4.9	262.681	779.38	24.816	1148.3	2795.0	1646.7	2.9095	5.9826	3.0731	1.28306	40.296
5.0	263.941	777.37	25.351	1154.6	2794.2	1639.6	2.9210	5.9737	3.0527	1.28639	39.446
5.1	265.181	775.37	25.888	1160.9	2793.4	1632.5	2.9323	5.9648	3.0325	1.28971	38.628
5.2	266.403	773.39	26.427	1167.0	2792.5	1625.5	2.9435	5.9561	3.0126	1.29302	37.840
5.3	267.608	771.42	26.968	1173.1	2791.6	1618.5	2.9546	5.9475	2.9930	1.29632	37.081
5.4	268.795	769.46	27.512	1179.1	2790.7	1611.5	2.9654	5.931	2.9736	1.29961	36.348
5.5	269.965	767.52	28.057	1185.1	2789.7	1604.6	2.9762	5.9307	2.9545	1.30290	35.642
5.6	271.120	765.59	28.605	1191.0	2788.7	1597.8	2.9868	5.9224	2.9356	1.30618	34.959
5.7	272.258	763.67	29.155	1196.8	2787.7	1590.9	2.9972	5.9142	2.9170	1.30946	34.300
5.8	273.382	761.77	29.707	1202.6	2786.7	1584.1	3.0075	5.9061	2.8985	1.31273	33.662
5.9	274.490	759.88	30.262	1208.3	2785.7	1577.4	3.0177	5.8981	2.8803	1.31600	33.045
6.0	275.585	758.00	30.818	1213.9	2784.6	1570.7	3.0278	5.8901	2.8623	1.31926	32.448
6.1	276.666	756.13	31.378	1219.5	2783.5	1564.0	3.0377	5.8823	2.8445	1.32253	31.870
6.2	277.733	75427	31.940	1225.1	2782.4	1557.3	3.0476	5.8745	2.8269	1.32579	31.309
6.3	278.787	752.42	32.504	1230.5	27812	1550.7	3.0573	5.8668	2.8095	1.32905	30.766
6.4	279.829	750.58	33.070	1236.0	2780.1	1544.1	3.0669	5.8592	2.7923	1.33230	30.238
6.5	280.858	748.75	33.640	1241.4	2778.9	1537.5	3.0764	5.8516	2.7752	1.33556	29.727
6.6	281.875	746.93	34.211	1246.7	2777.7	1530.9	3.0858	5.8441	2.7583	1.33882	29.230
6.7	282.880	745.11	34.786	1252.0	2776.4	1524.4	3.0951	5.8367	2.7416	1.34208	28.747
6.7	283.874	743.31	35.363	1257.3	27752	1517.9	3.1043	5.8293	2.7250	1.34533	28.278
6.9	284.857	741.51	35.943	1262.5	2773.9	1511.4	3.1134	5.8220	2.7086	1.34859	27.822
7.0	285.829	739.72	36.525	1267.7	2772.6	1505.0	3.1224	5.8148	2.6924	1.351 86	27.378
7.1	286.790	737.94	37.110	1272.8	2771.3	1498.5	3.1313	5.8076	2.6762	1.35512	26.947
7.2	287.741	736.17	37.698	1277.9	2770.0	1492.1	3.1402	5.8004	2.6603	1.35839	26.526
7.3	288.682	734.40	38.289	1282.9	2768.6	1485.7	3.1489	5.7933	2.6444	1.36166	26.117
7.4	289.614	732.64	38.883	1287.9	2767.3	1479.3	3.1576	5.7863	2.6287	1.36493	25.718

(*Continued*)

Saturated Steam (Pressure) (*Continued*)

$p,$ MPa	$T,$ °C	Density, kg/m³ ρ_L	ρ_V	Enthalpy, kJ/kg h_L	h_V	Δh	Entropy, kJ/(kg·K) S_L	s_V	Δ_s	Volume, cm³/g v_L	v_V
7.5	290.535	730.88	39.479	1292.9	2765.9	1473.0	3.1662	5.7793	2.6131	1.36821	25.330
7.6	291.448	729.14	40.079	1297.9	2764.5	1466.6	3.1747	5.7723	2.5976	1.37149	24.951
7.7	292.351	727.39	40.681	1302.8	2763.1	1460.3	3.1832	5.7654	2.5823	1.37477	24.581
7.8	293.245	725.66	41.287	1307.7	2761.6	1454.0	3.1915	5.7586	2.5671	1.37806	24.221
7.9	294.131	723.92	41.895	1312.5	27602	1447.7	3.1998	5.7518	2.5519	1.38136	23.869
8.0	295.008	72220	42.507	1317.3	2758.7	1441.4	3.2081	5.7450	2.5369	1.38467	23.526
8.1	295.876	720.47	43.122	1322.1	27572	1435.1	3.2162	5.7383	2.5220	1.38797	23.190
8.2	296.737	718.76	43.740	1326.8	2755.7	1428.8	3.2243	5.7316	2.5072	1.39129	22.863
8.3	297.589	717.04	44.361	1331.6	2754.1	1422.6	3.2324	5.7249	2.4925	1.39461	22.542
8.4	298.434	715.34	44.985	1336.3	2752.6	1416.3	3.2403	5.7183	2.4779	1.39795	22.229
8.5	299.271	713.63	45.613	1340.9	2751.0	1410.1	3.2483	5.7117	2.4634	1.40128	21.923
8.6	300.100	711.93	46.244	1345.6	2749.4	1403.9	3.2561	5.7051	2.4490	1.40463	21.624
8.7	300.922	71023	46.879	1350.2	2747.8	1397.7	3.2639	5.6986	2.4347	1.40799	21.332
8.8	301.737	708.54	47.517	1354.8	27462	1391.5	3.2717	5.6921	2.4204	1.41135	21.045
8.9	302.544	706.85	48.159	1359.3	2744.6	1385.3	3.2793	5.6856	2.4062	1.41473	20.765
9.0	303.345	705.16	48.804	1363.9	2742.9	1379.1	3.2870	5.6791	2.3922	1.41811	20.490
9.1	304.139	703.48	49.453	1368.4	2741.3	1372.9	3.2946	5.6727	2.3782	1.42151	20.221
9.2	304.926	701.80	50.105	1372.9	2739.6	1366.7	3.3021	5.6663	2.3642	1.42491	19.958
9.3	305.707	700.12	50.761	1377.4	2737.9	1360.5	3.3096	5.6599	2.3504	1.42833	19.700
9.4	306.481	698.44	51.421	1381.8	2736.2	1354.4	3.3170	5.6536	2.3366	1.43176	19.447
9.5	307.249	696.77	52.085	1386.2	2734.4	1348.2	3.3244	5.6473	2.3229	1.43520	19.199
9.6	308.010	695.09	52.753	1390.6	2732.7	1342.0	3.3317	5.64!0	2.3092	1.43865	18.956
9.7	308.766	693.42	53.424	1395.0	2730.9	1335.9	3.3390	5.6347	22957	1.44212	18.718
9.8	309.516	691.76	54.100	1399.4	2729.1	1329.7	3.3463	5.6284	22822	1.44560	18.484
9.9	310.259	690.09	54.779	1403.7	2727.3	1323.6	3.3535	5.6222	22687	1.44909	18.255
10.0	310.997	688.42	55.463	1408.1	2725.5	1317.4	3.3606	5.6160	22553	1.45259	18.030
10.2	312.456	685.10	56.843	1416.7	2721.8	1305.1	3.3749	5.6035	22287	1.45965	17.592
10.4	313.893	681.77	58.240	1425.2	2718.0	1292.8	3.3889	5.5912	22023	1.46676	17.170
10.6	315.308	678.45	59.655	1433.7	27142	1280.5	3.4028	5.5789	2.1761	1.47394	16.763
10.8	316.703	675.13	61.089	1442.1	2710.3	1268.2	3.4166	5.5667	2.1501	1.48119	16.370
11.0	318.079	671.81	62.541	1450.4	2706.3	1255.9	3.4303	5.5545	2.1242	1.48851	15.990
11.2	319.434	668.49	64.012	1458.7	2702.3	1243.6	3.4438	5.5423	2.0985	1.49590	15.622
11.4	320.771	665.17	65.504	1467.0	26982	1231.2	3.4572	5.5302	2.0730	1.50337	15.266
11.6	322.090	661.85	67.016	1475.2	2694.0	1218.8	3.4705	5.5181	2.0476	1.51093	14.922
11.8	323.391	658.52	68.550	1483.3	2689.8	1206.4	3.4836	5.5060	2.0224	1.51857	14.588

Superheated Steam

T, °C	1.5 MPa (Tₛ = 198.287°C) ρ, kg/ m³	h, kJ/ kg	s, kJ/ kg.K	1.5 MPa (Tₛ = 198.287°C) ρ, kg/ m³	h, kJ/ kg	s, kJ/ kg.K	2.0 MPa (Tₛ = 212.377°C) ρ, kg/ m³	h, kJ/ kg	s, kJ/ kg.K	2.5 MPa (Tₛ = 223.950°C) ρ, kg/ m³	h, kJ/ kg	s, kJ/ kg.K	T, °C
220	4.6087	2875.5	6.7934	7.1100	2850.2	6.5659	9.7870	2821.6	6.3867				220
230	4.4983	2898.4	6.8393	6.9191	2875.5	6.6166	9.4871	2850.2	6.4400	12.240	2821.8	6.2955	230
300	3.8762	3051.6	7.1246	5.8925	3038.2	6.9198	7.9677	3024.2	6.7684	10.107	3009.6	6.6459	300
400	3.2615	3264.5	7.4669	4.9256	3256.5	7.2710	6.6131	3248.3	7.1292	8.3251	3240.1	7.0170	400
500	2.8240	3479.1	7.7641	4.2524	3473.7	7.5718	5.6921	3468.2	7.4337	7.1433	3462.7	7.3254	500
600	2.4931	3698.6	8.0310	3.7484	3694.7	7.8405	5.0097	3690.7	7.7043	6.2769	3686.8	7.5979	600
700	2.2330	3924.1	8.2755	3.3544	3921.1	8.0860	4.4790	3918.2	7.9509	5.6070	3915.2	7.8455	700
800	2.0227	4156.1	8.5024	3.3068	4153.8	8.3135	4.0528	4251.5	8.1790	5.0706	4149.2	8.0743	800
900	1.8490	4394.8	8.7150	2.7750	4392.9	8.5266	3.7021	4391.1	8.3925	4.6302	4389.3	8.2882	900
1000	1.7030	4639.9	8.9155	2.5553	4638.5	8.7274	3.4081	4637.0	8.5936	4.2615	4635.6	8.4896	1000
1200	1.4710	5148.9	9.2866	2.2065	5147.9	9.0988	2.9421	5147.0	8.9654	3.6778	5146.0	8.8618	1200

T, °C	3.0 MPa (Tₛ = 233.853°C) ρ, kg/ m³	h, kJ/ kg	s, kJ/ kg.K	3.5 MPa (Tₛ = 242.557°C) ρ, kg/ m³	h, kJ/ kg	s, kJ/ kg.K	4.0 MPa (Tₛ = 250.354°C) ρ, kg/ m³	h, kJ/ kg	s, kJ/ kg.K	4.5 MPa (Tₛ = 257.437°C) ρ, kg/ m³	h, kJ/ kg	s, kJ/ kg.K	T, °C
250	14.159	2856.5	6.2893	17.019	2829.7	6.1764							250
260	13.718	2886.4	6.3459	16.424	2862.9	6.2393	19.314	2837.1	6.1383	22.435	2808.6	6.0397	260
300	12.318	2994.3	6.5412	14.609	2978.4	6.4484	16.987	2961.7	6.3639	19.464	2944.2	6.2854	300
400	10.042	3231.7	6.9234	11.826	3223.2	6.8427	13.618	3214.5	6.7714	15.4390	3205.6	6.7070	400
500	8.6062	3457.2	7.2359	10.081	3451.6	7.1593	11.568	3446.0	7.0922	13.0680	3440.4	7.0323	500
600	7.5503	3682.2	7.5103	8.8297	3678.9	7.4356	10.115	3674.9	7.3705	11.4070	3670.9	7.3127	600
700	6.7383	3912.2	7.7590	7.8729	3909.3	7.6854	9.0109	3906.3	7.6214	10.152	3903.3	7.5646	700
800	6.0903	4146.9	7.9885	7.1118	4144.6	7.9156	8.1352	4142.3	7.8523	9.1605	4140.0	7.7962	800
900	5.5593	4387.5000	8.2028	6.4895	4385.7	8.1303	7.4026	4389.9	8.0674	9.3528	4382.1	8.0118	900
1000	5.1154	4634.1000	8.4045	5.9698	4632.7	8.3324	6.8248	4631.2	8.2697	7.6803	4629.8	8.2144	1000
1200	4.4136	5145.0000	8.7770	5.1494	5144.1	8.7053	5.8852	5143.1	8.6430	6.6211	5142.2	8.5880	1200

T, °C	5.0 MPa (Tₛ = 263.941°C) ρ, kg/ m³	h, kJ/ kg	s, kJ/ kg.K	6.0 MPa (Tₛ = 275.585°C) ρ, kg/ m³	h, kJ/ kg	s, kJ/ kg.K	7.0 MPa (Tₛ = 285.829°C) ρ, kg/ m³	h, kJ/ kg	s, kJ/ kg.K	8.0 MPa (Tₛ = 295.008°C) ρ, kg/ m³	h, kJ/ kg	s, kJ/ kg.K	T, °C
280	23.655	2858.1	6.0909	30.121	2805.3	2805.3							280
290	22.802	2893.0	6.1536	28.767	2847.5	6.0034	35.659	2794.1	5.8529				290
300	22.053	2925.7	6.2110	27.632	2885.5	6.0703	33.907	2839.9	5.9337	41.188	2786.5	5.7937	300
400	17.290	3196.7	6.6483	21.088	3178.2	6.5432	25.026	3159.2	6.4502	29.117	3139.4	6.3658	400
500	14.581	3434.7	6.9781	17.646	3423.1	6.8826	20.765	3411.4	6.8000	23.942	3399.5	6.7266	500
600	12.706	3666.8	7.2605	15.322	3658.7	7.1693	17.965	3650.6	7.0910	20.634	3642.4	7.0221	600

(Continued)

	5.0 MPa (T_s = 263.941 °C)			6.0 MPa (T_s = 275.585 °C)			7.0 MPa (T_s = 285.829 °C)			8.0 MPa (T_s = 295.008 °C)			
T, °C	ρ, kg/ m³	h, kJ/ kg	s, kJ/ kg.K	ρ, kg/ m³	h, kJ/ kg	s, kJ/ kg.K	ρ, kg/ m³	h, kJ/ kg	s, kJ/ kg.K	ρ, kg/ m³	h, kJ/ kg	s, kJ/ kg.K	T, °C
700	11.297	3900.3	7.5136	13.597	3894.3	7.4246	15.911	3888.2	7.3486	18.239	3882.2	7.2821	700
800	10.188	4137.7	7.7458	12.248	4133.1	7.6582	14.315	4128.4	7.5836	16.390	4123.8	7.5184	800
900	9.2861	4380.2	7.9618	11.156	4376.6	7.8751	13.029	4373.0	7.8014	14.907	4369.3	7.7371	900
1000	8.5364	4628.3	8.1648	10.250	4625.4	8.0786	11.966	4622.5	8.0055	13.684	4619.6	7.9419	1000
1200	7.3571	5141.2	8.5388	8.8291	5139.3	8.4534	10.301	5137.4	8.3810	11.774	5135.5	8.3181	1200

Compressed Water

T, °C	0.10 MPa (T_s = 99.606°C) ρ	h	S	0.50 MPa (T_s = 151.831°C) ρ	h	s	1.0 MPa (T_s = 179.878°C) ρ	h	s	1.5 MPa (T_s = 198.287°C) ρ	h	s	T, °C
t_s (L)	958.63	147.50	1.3028	915.29	640.09	1.8604	887.13	762.52	2.1381	866.65	844.56	2.3143	t_s (L)
t_s (V)	059034	2674.9	7.3588	2.6680	2748.1	6.8207	5.1450	2777.1	6.5850	7.5924	2791.0	6.4430	t_s (V)
0	999.84	0.06	−0.00015	1000.05	0.47	−0.00012	1000.30	0.98	−0.00009	1000.55	1.48	−0.00006	0
5	999.97	21.12	0.07625	1000.16	21.52	0.07625	1000.41	22.01	0.07624	1000.65	22.51	0.07623	5
10	999.70	42.12	0.15108	999.89	42.51	0.15104	1000.13	42.99	0.15100	1000.37	43.48	0.15095	10
15	999.10	63.08	0.22445	999.29	63.46	0.22439	999.52	63.94	0.22431	999.75	64.41	0.22424	15
20	998.21	84.01	0.29646	998.39	84.38	0.29638	998.62	84.85	0.29628	998.85	85.32	0.29617	20
25	997.05	104.92	0.36720	997.23	105.29	0.36710	997.45	105.75	0.36697	997.68	106.21	0.36684	25
30	995.65	125.82	0.43673	995.83	126.19	0.43660	996.05	126.64	0.43645	996.27	127.10	0.43630	30
35	994.03	146.72	0.50510	994.21	147.08	0.50496	994.43	147.53	0.50478	994.65	147.98	0.50441	35
40	992.22	167.62	0.57237	992.39	167.97	0.57221	992.61	168.41	0.57202	992.83	168.86	0.57182	40
45	990.21	188.51	0.63858	990.39	188.86	0.63840	990.61	189.30	0.63819	990.82	189.74	0.63798	45
50	988.03	209.42	0.70377	988.21	209.76	0.70358	988.43	210.19	0.70335	988.64	210.62	0.70312	50
55	985.69	230.33	0.76798	985.87	230.67	0.76778	986.09	231.09	0.76753	986.30	231.52	0.76728	55
60	983.20	251.25	0.83125	983.37	251.58	0.83104	983.59	252.00	0.83077	983.81	252.42	0.83051	60
65	980.55	272.18	0.89361	980.73	272.51	0.89338	980.95	272.92	0.89310	981.16	273.34	0.89282	65
70	977.76	293.12	0.95509	977.94	293.45	0.95485	978.16	293.86	0.95455	978.38	294.27	0.95426	70
75	974.84	314.08	1.0157	975.02	314.40	1.0155	975.24	314.81	1.0152	975.46	315.21	1.0148	75
80	971.79	335.05	1.0755	971.97	335.37	1.0753	972.19	335.77	1.0750	472.42	336.17	1.0746	80
85	968.61	356.05	1.1346	968.79	356.36	1.1343	969.02	356.75	1.1340	969.24	357.15	1.1336	85
90	965.31	377.06	1.1928	965.49	377.37	1.1926	965.72	377.76	1.922	965.95	378.15	1.1918	90
95	961.89	398.10	1.2504	962.07	398.41	1.2501	962.30	398.79	1.2497	962.53	399.17	1.2493	95
100	ρ (kg/m³)			958.54	419.47	1.3069	958.77	419.84	1.3065	959.00	420.22	1.3061	100
105	h (kJ/kg)			954.88	440.55	1.3630	955.12	440.92	1.3626	955.36	441.29	1.3622	105
110	s (kJ/kg.K)			951.12	461.67	1.4185	951.36	462.04	1.4181	951.60	462.40	1.4177	110
115				947.24	482.83	1.4734	947.49	483.19	1.4729	947.74	483.55	14725	115
120				943.26	504.02	1.5276	943.51	504.38	1.5272	943.76	504.73	1.5267	120
125				939.16	525.26	1.5813	939.42	525.60	1.5808	939.67	525.95	1.5809	125
130				934.95	456.54	1.6344	935.21	546.88	1.6339	935.47	547.22	1.6334	130
135				930.64	567.84	1.6870	930.90	568.20	1.6865	931.17	568.53	1.6860	135
140				926.21	589.25	1.7391	926.48	589.58	1.6865	926.75	589.90	1.7380	140
145				921.67	610.69	1.7907	921.95	611.01	1.7901	922.23	611.33	1.7896	145
150				917.02	632.19	1.8418	917.31	632.50	1.8412	917.65	632.87	1.8405	150
155							912.55	654.06	1.8919	912.90	654.42	1.8912	155
160							907.68	675.70	1.9421	908.04	676.05	1.8912	160
165							902.69	697.41	1.9919	903.06	697.74	1.9912	165
170							897.58	719.20	2.0414	897.96	719.52	2.40406	170
175							892.35	741.08	2.0905	892.74	741.39	2.0897	175
180										887.40	763.35	2.1384	180
185										881.40	785.42	2.1868	185
190										876.32	807.59	2.2350	190
195										870.58	829.88	2.2828	195

	2.0 MPa (T_s = 212.377°C)			4.0 MPa (T_s = 250.354°C)			6.0 MPa (T_s = 275.585°C)			8.0 MPa (T_s = 295.008°C)			
T, °C	ρ	h	s	ρ	h	s	ρ	h	s	ρ	h	s	T, °C
t_s (L)	849.80	908.50	2.4468	798.37	1087.5	2.7968	758.00	1213.9	3.0278	722.20	1317.3	3.2081	t_s (L)
t_s (V)	10.042	2798.3	6.3390	20.090	2800.8	6.0696	30.818	2784.6	5.8901	42.507	2758.7	5.7450	t_s (V)
0	1000.81	1.99	−0.00003	1001.82	4.02	0.00009	1002.82	6.04	0.00019	1003.82	8.06	0.00027	0
5	1000.90	23.01	0.07622	1001.88	24.99	0.07617	1002.85	26.97	0.07611	1003.82	28.94	0.07603	5
10	1000.61	43.97	0.15091	1001.56	45.91	0.15072	1002.50	47.85	0.15052	1003.45	49.79	0.15031	10
15	999.99	64.89	0.22416	1000.92	66.80	0.22385	1001.84	68.71	0.22353	1002.76	70.61	0.22320	15
20	999.08	85.79	0.29607	999.99	87.67	0.29564	1000.89	89.54	0.29522	1001.80	91.41	0.29478	20
25	997.90	106.68	0.36671	998.80	108.52	0.36619	999.69	110.37	0.36566	1000.58	112.21	0.36513	25
30	996.49	127.55	0.43615	997.38	129.37	0.43553	998.26	131.19	0.43492	999.14	133.01	0.43430	30
35	994.87	148.43	0.50444	995.75	150.22	0.50374	996.62	152.01	0.50304	997.49	153.80	0.50234	35
40	993.05	169.30	0.57163	993.92	171.07	0.57085	994.79	172.84	0.57007	995.65	174.60	0.56929	40
45	991.04	190.17	0.63776	991.91	191.92	0.63691	992.78	193.66	0.63606	993.64	195.41	0.63521	45
50	988.86	211.06	0.70289	989.73	212.78	0.70196	990.59	214.50	0.70104	991.45	216.22	0.70012	50
55	986.52	231.94	0.76704	987.39	233.64	0.76604	988.25	235.34	0.76505	989.11	237.04	0.76406	55
60	984.02	252.84	0.83024	984.89	254.52	0.82918	985.76	256.20	0.82812	986.62	257.88	0.82707	60
65	981.38	273.75	0.89254	982.26	275.41	0.89141	983.12	277.07	0.89029	983.99	278.72	0.88917	65
70	978.60	294.68	0.95396	979.48	296.31	0.95277	980.35	297.95	0.95159	981.22	299.58	0.95041	70
75	975.69	315.61	1.0145	976.57	317.23	1.0133	977.45	318.84	1.0120	978.32	320.45	1.0108	75
80	972.64	336.57	1.0743	973.53	338.16	1.0730	974.42	339.75	1.0717	975.30	341.34	1.0704	80
85	969.47	357.54	1.1333	970.37	359.11	1.1319	971.26	360.68	1.1305	972.15	362.25	1.1292	85
90	966.18	378.53	1.1915	967.09	380.08	1.1900	967.99	381.63	1.1886	968.89	383.18	1.1872	90
95	962.77	399.55	1.2490	963.69	401.08	1.2475	964.60	402.60	1.2460	965.51	404.13	1.2445	95
100	959.24	420.59	1.3057	960.17	422.10	1.3042	961.10	423.60	1.3026	962.02	425.11	1.3011	100
105	955.60	441.66	1.3618	956.54	443.15	1.3602	957.48	444.63	1.3586	958.42	446.11	1.3570	105
110	951.84	462.77	1.4173	952.80	464.22	1.4156	953.76	465.68	1.4139	954.71	467.15	1.4123	110
115	947.98	483.90	1.4721	948.96	485.34	1.4703	949.93	486.77	1.4686	950.89	488.21	1.4669	115
120	944.01	505.08	1.5263	945.00	506.49	1.5245	945.99	507.90	1.5227	946.97	509.31	1.5209	120
125	939.92	526.29	1.5799	940.93	527.68	1.5780	941.94	529.06	1.5762	942.94	530.45	1.5743	125
130	935.73	547.55	1.6330	936.76	548.91	1.6310	937.79	550.27	1.6291	938.80	551.63	1.6272	130
135	931.43	568.86	1.6855	932.48	570.19	1.6835	933.53	571.53	1.6815	934.57	572.86	1.6795	135
140	927.02	590.22	1.7375	928.10	591.53	1.7354	929.16	592.83	1.7334	930.22	594.14	1.7313	140
145	922.50	611.64	1.7890	923.60	612.91	1.7869	924.69	614.19	1.7848	925.78	615.47	1.7827	145
150	917.87	633.12	1.8401	919.00	634.36	1.8379	920.11	635.61	1.8357	921.22	636.86	1.8335	150
155	913.13	654.67	1.8907	914.28	655.87	1.8884	915.43	657.09	1.8862	916.56	658.30	1.8839	155
160	908.27	676.28	1.9409	909.46	677.45	1.9385	910.63	678.63	1.9362	911.79	679.82	1.9339	160
165	903.30	697.97	1.9907	904.52	699.11	1.9882	905.72	700.25	1.9858	906.92	701.40	1.9834	165
170	898.21	719.74	2.0401	899.46	720.84	2.0376	900.70	721.95	2.0351	901.93	723.06	2.0326	170
175	893.00	741.60	2.0892	894.29	742.66	2.0865	895.56	743.73	2.0839	896.82	744.80	2.0813	175
180	887.67	763.56	2.1379	888.99	764.57	2.1352	890.30	765.60	2.1325	891.60	766.63	2.1298	180
185	882.20	785.61	2.1863	883.57	786.58	2.1835	884.92	787.56	2.1807	886.26	788.55	2.1779	185
190	876.61	807.77	2.2344	878.02	808.69	2.2315	879.41	809.63	2.2286	880.79	810.57	2.2257	190
195	870.87	830.05	2.2822	872.33	830.92	2.2792	873.78	831.80	2.2762	875.20	832.70	2.2732	195
200	865.00	852.45	2.3298	866.51	853.27	2.3267	868.00	854.09	2.3235	869.48	854.94	2.3205	200
210	852.79	897.66	2.4244	854.42	898.35	2.4210	856.03	899.06	2.4176	857.61	899.79	2.4143	210

(*Continued*)

T, °C	2.0 MPa (T_s = 212.377°C)			4.0 MPa (T_s = 250.354°C)			6.0 MPa (T_s = 275.585°C)			8.0 MPa (T_s = 295.008°C)			T, °C
	ρ	h	s	ρ	h	s	ρ	h	s	ρ	h	s	
220	ρ (kg/m³)			841.70	944.04	2.5146	843.44	944.61	2.5109	845.15	945.20	2.5073	220
230	h (kJ/kg)			828.28	990.42	2.6077	830.17	990.82	2.6037	832.03	991.25	2.5997	230
240	s (kJ/kg.K)			814.06	1037.6	2.7005	816.14	1037.8	2.6961	818.18	1038.1	2.6919	240
250				798.92	1085.8	2.7935	801.23	1085.7	2.7886	803.49	1085.7	2.7839	250
260							785.32	1134.7	2.8814	787.84	1134.5	2.8761	260
270							768.19	1185.1	2.9750	771.05	1184.5	2.9690	270
280										752.88	1236.0	3.0631	**280**
290										732.98	1289.6	3.1590	**290**

Source: Thermodynamic Properties of Water: Tabulation from the IAPWS Formulation 1995 for the Thermodynamic Properties of Ordinary Water Substance for General and Scientific use.

A.6   The Moody Chart

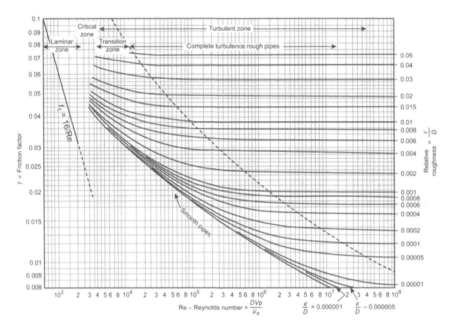

Source: L. F. Moody, *Friction Factors for Pipe flow*, vol. 66, pp. 671–684, 1944. ASME Trans.

Recommended Roughness, ε

Material	Condition	ft	mm	Uncertainty, %
Steel	Sheet metal, new	0.00016	0.05	± 60
	Stainless, new	0.000007	0.002	± 50
	Commercial, new	0.00015	0.046	± 30
	Riveted	0.01	3.0	± 70
	Rusted	0.007	2.0	± 50
Iron	Cast, new	0.00085	0.26	± 50
	Wrought, new	0.00015	0.046	± 20
	Galvanized, new	0.0005	0.15	± 40
	Asphalted cast	0.0004	0.12	± 50
Brass	Drawn, new	0.000007	0.002	± 50
Plastic	Drawn tubing	0.000005	0.0015	± 60
Glass	—	Smooth	Smooth	
Concrete	Smoothed	0.00013	0.04	± 60
	Rough	0.007	2.0	± 50
Rubber	Smoothed	0.000033	0.01	± 60
Wood	Stave	0.0016	0.5	± 40
Copper	—	0.000005	0.0015	

A.7 Minor Resistance Coefficient and Equivalent Length

(a) Minor Resistance Coefficient, $K = K_L$, D = ND (mm)

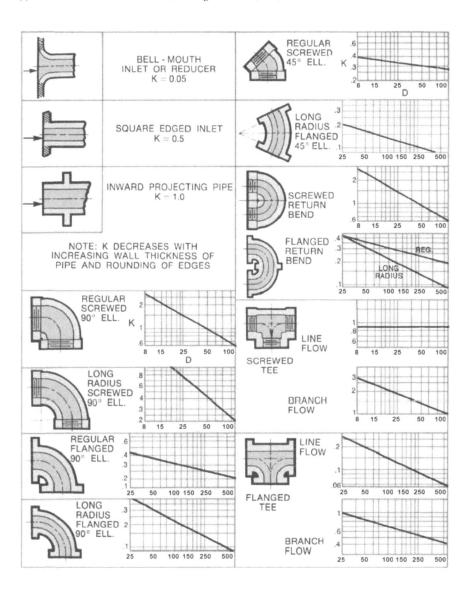

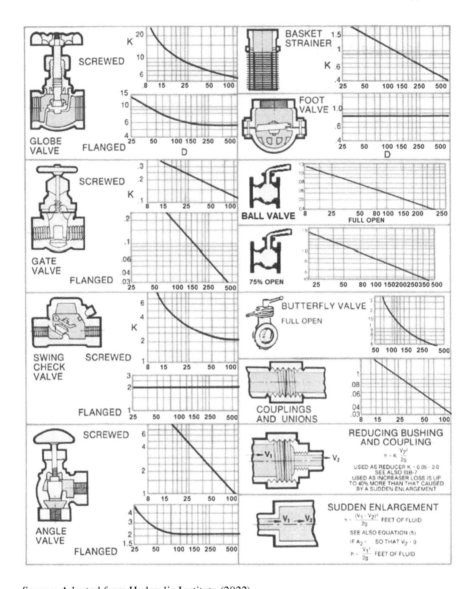

Source: Adapted from Hydraulic Institute (2022).

(b) Equivalent Length of Valves and Fittings, L_{eq} (m)

Screwed/Flanged Fittings		Pipe Size (mm)								
		15	20	25	32	40	50	65	80	100
Elbows	Regular 90°	1.1/0.3	1.3/0.4	1.6/0.5	2.0/0.6	2.3/0.7	2.6/0.9	2.8/1.1	3.4/1.3	4.0/1.8
	Long radius 90°	0.7/0.3	0.7/0.4	0.8/0.5	1.0/0.6	1.0/0.7	1.1/0.8	1.1/0.9	1.2/1.0	1.4/1.3
	Regular 45°	0.2/0.1	0.3/0.2	0.4/0.6	0.5/0.3	0.6/0.4	0.8/0.5	1.0/0.6	1.2/0.8	1.7/1.1
Tees	Line flow	0.5/0.2	0.7/0.3	1.0/0.3	1.4/0.4	1.7/0.5	2.3/0.5	2.8/0.6	3.7/0.7	5.2/0.9
	Branch flow	1.3/0.6	1.6/0.8	2.0/1.0	2.7/1.3	3.0/1.6	3.7/2.0	4.0/2.3	5.2/2.9	6.4/3.7
Return bends	Regular 180°	1.1/0.3	1.3/0.4	1.6/0.5	2.0/0.6	24.1/0.7	33.6/0.9	2.8/1.1	3.4/1.3	4.0/1.8
Valves	Globe	6.7/11.6	7.3/12.2	8.8 13.7	11.3/16.5	12.8/18.0	16.5/21.4	18.9/23.5	24.1/28.7	33.6/36.6
	Gate	0.2/0.0	0.2/0.0	0.3/0.0	0.3/0.0	0.4/0.0	0.5/0.8	0.5/0.9	0.6/1.0	0.8/1.3
	Angle	4.6/4.6	4.6/4.6	5.2/5.2	5.5/5.5	5.5/5.5	5.5/6.4	5.5/6.7	5.5/8.5	5.5/11.6
	Swing check	2.4	2.7	3.4	4.0	4.6	5.8	6.7	8.2	11.6
Strainer		1.5	2.0	2.3	5.5	6.1	8.2	8.8	10.4	12.8

A.8 Hydraulic Diameters of Regular Shapes

Shape		Area (A)	Perimeter (P)	Hydraulic Diameter (D_h)
Circular		$\dfrac{\pi D^2}{4}$	πD	$\dfrac{4(\pi D^2/4)}{\pi D} = D$
Square		a^2	$4a$	$\dfrac{4a^2}{4a} = a$
Rectangular		ab	$2(a+b)$	$\dfrac{4ab}{2(a+b)} = \dfrac{2ab}{a+b}$
Triangular		$\dfrac{\sqrt{3}}{4}a^2$	$3a$	$D_h = \dfrac{\sqrt{3}}{3}a$

Shape		Area (A)	Perimeter (P)	Hydraulic Diameter (D_h)
Circular		$(\theta - \sin\theta)\dfrac{D^2}{8}$	$\dfrac{D}{2}\theta$	$D\left(1 - \dfrac{\sin\theta}{\theta}\right)$
Trapezoidal		$y(b+zy)$	$b + 2y\sqrt{1+z^2}$	$\dfrac{4y(b+2y)}{b+2y\sqrt{(1+z^2)}}$
Rectangular		by	$b + 2y$	$\dfrac{4by}{b+2y}$
Triangular		zy^2	$2y\sqrt{1+z^2}$	$\dfrac{2zy}{y\sqrt{(1+z^2)}}$

Bibliography

A Group of Authorities, *Marine Engineering*. The Society of Naval Architects and Marine Engineers, 1992.

American Society of Heating, Refrigerating and Air-Conditioning Engineers (ASHRAE), *Understanding Psychrometrics*. American Society of Heating, Refrigerating, and Air-Conditioning Engineers, 3rd edition, 2013.

American Society of Heating, Refrigerating and Air-Conditioning Engineers (ASHRAE), *Handbook of Fundamentals*. The American Society of Heating, Refrigerating, and Air-Conditioning Engineers, 2017.

American Society of Heating, Refrigerating and Air-Conditioning Engineers (ASHRAE), *Handbook of HVAC Application*. The American Society of Heating, Refrigerating, and Air-Conditioning Engineers, 2019.

American Society of Mechanical Engineers (ASME), I. Power Boilers, IV. Heating Boilers in *Boiler and Pressure Vessel Code (BPVC)*, Y14.5, ASME, 2009.

Babcock & Wilcox Company, *Steam: Its Generation and Use*. The Babcock & Wilcox Company, 2005.

Bearg, D. W. *Indoor Air Quality and HVAC Systems*. CRC Press, 1993.

Bejan, A. *Advanced Engineering Thermodynamics*. Wiley, 4th edition, 2016.

Bowman, R. A., Mueller, A. C., Nagle, W. M. *Mean Temperature Difference in Design*. Transactions of ASME 62, p. 283, 1940.

Carrier Air Conditioning Company, *Handbook of Air Conditioning System Design*. McGraw-Hill, 1965.

Cengel, Y. A., Boles, M. A. *Thermodynamics – An Engineering Approach*. McGraw-Hill, 9th edition, 2019.

Cengel, Y. A., Cimbala, J. M. *Fluid Mechanics – Fundamentals & Applications*. McGraw-Hill, 4th edition, 2018.

Cengel, Y. A., Ghajar, A. J. *Heat and Mass Transfer – Fundamentals & Applications*. McGraw-Hill, 6th edition, 2020.

Churchill, S. W. *Friction Factor Equation Spans All Fluid-Flow Regimes*. Chemical Engineering 84, pp. 91–92, 1977.

Clausius, R. *On the Moving Force of Heat, and the Laws Regarding the Nature of Heat Itself Which are Deducible Therefrom*. The London, Edinburgh, and Dublin Philosophical Magazine and Journal of Science 2, pp. 1–21, 1851.

Clausius, R. On a Modified Form of the Second Fundamental Theorem in the Mechanical Theory of Heat. The London, Edinburgh, and Dublin Philosophical Magazine and Journal of Science 12 (4), 1856.

Colebrook, C. F. *Turbulent Flow in Pipes, with Particular Reference to the Transition between the Smooth and Rough Pipe Laws*. Journal of the Institution of Civil Engineers 11 (4), pp. 133–156, 1939.

Duffie, J. A., Beckman, W. A. *Solar Engineering of Thermal Processes*. Wiley, 3rd edition, 2006.

F-Chart Software, "EES," 2022. https://fchartsoftware.com/

Fisher Controls International Inc., *Control Valve Handbook*. Fisher Controls International Inc., 3rd edition. 2001.

Gatley, D. P. Psychrometric Chart Celebrates 100th Anniversary. ASHRAE Journal, 46 (11), pp. 16–20, 2004.

Grimm, N. R., Rosaler, R. C. *HVAC Systems and Components Handbook.* McGraw-Hill, 2nd edition, 1997.

Gu, Y. *Absorption Chiller With Cooling and Heating for Energy Saving and Environment Friendly.* International Journal of Engineering and Innovative Technology (IJEIT) 7 (11), 2018.

Gu, Y. *An Excellent SCCHP Design Project in Thermal-Fluids Education.* ASTFE Digital Library 2379-1748, pp. 603–612, 2020.

Gu, Y. *Analysis of Air Cooling and Dehumidification Process Through Cooling Coils.* ASTFE Digital Library 2379-1748, pp. 1327–1334, 2021.

Gu, Y. *A Study of Cogeneration Combined Steam Turbine Power Plant With Absorption Chiller of HVAC.* ASTFE Digital Library 2379-1748, pp. 273–282, 2022.

Haaland, S. E. *Simple and Explicit Formulas for the Friction Factor in Turbulent Pipe Flow.* Journal of Fluids Engineering 105 (1), pp. 89–90, 1983.

Hardee, R. T., Sines, J. L. *Piping System Fundamentals.* Engineered Software Inc. Press, 2nd edition, 2012.

Jenkins, N., Ekanayake, J. *Renewable Energy Engineering.* Cambridge University Press, 1st edition, 2017.

Jones, J. B. *Engineering Thermodynamics – An Introductory Textbook.* Wiley, 2nd edition, 1986.

Kanoglu, M., Cengel, Y. A., Cimbala, J. M. *Fundamentals and Application of Renewable Energy.* McGraw-Hill, 2020.

Kern, D. Q., Kraus, A. D. *Extended Surface Heat Transfer.* McGraw-Hill, 1972.

Khartchenko, N. V. *Green Power: Eco-Friendly Energy Engineering.* Delhi: Tech Books, 2004.

Khartchenko, N. V., Kharchenko, V. M. *Advanced Energy Systems.* CRC Press, 2nd edition, 2014.

Linderburg, M. R. *Core Engineering Concepts for Students and Professionals.* Professional Publications, 2010.

Mitchell, J. W. *Energy Engineering.* Wiley, 1983.

Moody, L. F. Friction Factors for Pipe Flows. Transactions of the ASME 66 (8), pp. 671–678, 1944.

Moran, M. J., Shapiro, H. N., Boettner, D. D., Bailey, M. B. *Fundamentals of Engineering Thermodynamics.* John Wiley & Sons, 9th edition, 2018.

Munters Cargocaire, *The Dehumidification Handbook.* Munters Cargocaire, 2nd edition, 1990.

Nayyar, M. L. *Piping Handbook.* McGraw-Hill, 7th edition, 2000.

Pan, L., Shi, W. Investigation on the Pinch Point Position in Heat Exchangers. Journal of Thermal Science 25, pp. 258–265, 2016.

Pis'mennyi, E., Polupan, G., Mariscal, I. C., Silva, F. S., Pioro, I. *Handbook for Transversely Finned Tube Heat Exchanger Design.* Academic Press, 2016.

Planck, M. *Treatise on Thermodynamics.* Translated by Ogg, A. London: Longmans Green, 1897.

Ramsey, M. A. *Tested Solutions to: Design Problems in Air Conditioning and Refrigeration.* Industrial Press, 1966.

Reynolds, O. On the Experimental Investigation of the Circumstances which Determine Whether the *Motion* of Water Shall be Direct or Sinuous, and the Law of Resistance in Parallel Channels. *Philosophical Transactions of the Royal Society of London,* Vol. 174, (1883), pp. 935–982.

Schaschke, C. J. *Solved Practical Problems in Fluid Mechanics.* Taylor & Francis, 2016.

Schlichting, H. *Boundary Layer Theory.* McGraw-Hill, 1979.

Sherman, M. H. (Ed.). *Air Change Rate and Airtightness in Buildings.* ASTM, 1990

Stasiulevicius, J., Skrinska, A. *Heat Transfer of Finned Tube Bundles in Crossflow.* Hemisphere Publishing Corporation, 1987.

Stevens, R. A., Fernanders, J., Woolf, J. R. Mean *Temperature* Difference in One, Two, and Three Pass Crossflow Heat Exchangers. Transactions of the ASME 79 (2), pp. 287–296, 1957.

Swanee, P. K., Jain, A. K. Explicit Equations for Pipe-Flow Problems. *Journal of the Hydraulics Division* 102 (5), 1976.

Taborek, J., Hewitt, G. F., Afgan, N. *Heat Exchangers: Theory and Practice.* Hemisphere Publishing Corporation, 1983.

Vedavarz, A., Kumar, S., Hussain, M. I. *The Handbook of Heating, Ventilation and Air Conditioning (HVAC) for Design and Implementation.* Industrial Press Inc., 2007.

White, F. M. *Fluid Mechanics.* McGraw-Hill, 7th edition, 2011.

Zukauskas, A., Ulinskas, R. *Efficiency Parameters for Heat Transfer in Tube Banks.* Heat Transfer Engineering 6, pp. 19–25, 1985.

Zukauskas, A. *Heat Transfer from Tubes in Cross Flow.* In *Handbook of Single Phase Convective Heat Transfer*, Kakac, S., Shah, R. K., Aung, W. (Eds.). Wiley Interscience, 1987.

Index

For Product Safety Concerns and Information please contact our
EU representative GPSR@taylorandfrancis.com Taylor & Francis
Verlag GmbH, Kaufingerstraße 24, 80331 München, Germany